JANNER'S COMPLETE BUSINESS LETTERWRITER

FIFTH EDITION

For
Isabel, Esther and Phoebe
Tali, Natan and Ella Avigail
with my love

Janner's Complete Business Letterwriter
Fifth edition

Lord Janner of Braunstone QC
with Matthew Solon

Gower

First published in 1970 as *The Businessman's Guide to Letter-writing and to the Law on Letters*
Second edition 1977
Third edition 1983
Fourth edition 1989

This edition published by
Gower Publishing Limited
Gower House
Croft Road
Aldershot
Hampshire GU11 3HR
England

Gower
Old Post Road
Brookfield
Vermont 05036
USA

British Library Cataloguing in Publication Data
Janner, Greville, 1928–
    Janner's complete business letterwriter. – 5th ed.
    1. Commercial correspondence 2. English language – Written English
    I. Title II. Solon, Matthew III. Complete business
    letterwriter IV. Janner's complete letterwriter V. Complete
    letterwriter
    808'.066'65

ISBN 0 566 07963 1

Library of Congress Cataloging-in-Publication Data
Janner, Greville.
    [Complete business letterwriter]
    Janner's complete business letterwriter / Lord Janner of Braunstone with Matthew Solon. – 5th ed.
        p. cm..
    Rev. ed of: The businessman's guide to letter-writing and to the law on letters.
    Includes index.
    ISBN 0-566-07963-1
    1. Commercial correspondence. 2. Commercial law. I. Solon, Matthew. II. Mitchell, Ewan. Businessman's guide to letter-writing and to the law on letters. III. Title.
    HF5721.J33 1998
    808', 066651—dc21   98–6782
                                                                                              CIP

Typeset in 10 point Palatino by Wearset, Boldon, Tyne & Wear
and printed in Great Britain at the University Press, Cambridge.

Contents

Preface

The essence of fine letterwriting remains unchanged. Style –
through appearance and language. Impact – through simplicity
and structure. Targeting your readers; what they want; and
your message.

Still, language and techniques advance. And the book which
I was so proud of when the previous edition appeared in 1989
no longer pleased me. So I have updated, rewritten, repolished,
revised – and I hope that the result will help you to get the
results that you want from your letters, whatever they may be.

Sam Goldwyn once told an author whose work he was given
to read: 'I read part of it all the way through.' I have read all of
this book, all the way through, and after 28 years and four edi-
tions I was amazed at the permanence of the good rules. But
techniques have changed and, with the help of Matthew Solon,
I have recorded those changes.

As a lawyer, I have enjoyed Will Rogers' dictum: 'The
minute that you read something that you can't understand, you
can almost be sure it was drawn up by a lawyer.' I hope that I
have translated the inexplicable into the understandable and
breathed some life and enjoyment into what could be a dull
subject. And that includes the chapters on the law on letters

and in litigation. But as Goldwyn was right, and books of this sort are not only for dipping into and for reference but also for relaxed reading, I have kept the chapters short.

So I hope that this book will continue to be the definitive guide to the business letterwriter's art. It is the product of decades of practical letterwriting – professional and personal, organizational and social, legal and general. I hope it will give you pleasure and help.

Finally, my thanks to Matthew Solon; my daughter Laura; to Paul Secher and David Roth; and to my patient secretaries, Pat Garner and Margaret Lancaster, for all their help in preparing this new edition.

Greville Janner
Lord Janner of Braunstone QC

For details of training in the skills of business letterwriting, contact: Paul Secher LLB, JSB Training, Dove House, Arcadia Avenue, London N3 2JU. Telephone 0181-371 7000; Fax 0181-371 7001.

Part I

Form and Formalities

1

The shape of a letter

A fine letter must have form and shape. For ordinary, simple notes – no problem. By the time you have begun with 'Thank you for your letter . . .' and ended with 'Best wishes', the middle will have looked after itself. But when a letter is important; when it matters to you; when you are prepared to take time to work on the words – then you should divide your draft into three parts: the opening, the body and the closing. Create the skeleton in note form; clothe it with the ideas you want to put across; and so the letter takes shape. As with a structured speech or any other presentation, say what you are going to say; say it; then say what you've said.

The first and last sentences are of the utmost importance. Skip the traditional opening gambit ('We are in receipt of . . .'). Start with a clear first sentence, to catch the reader's interest, to sum up what is to come, and to make sure that your words will be taken seriously.

Next, each idea should lead on logically to the following one. Just as each bone of the human body is attached to its fellow, so the ideas in a letter should be jointed. The flow of ideas needs rhythm. Disjointed ideas, dislocated thoughts, fractured theories, these are the hallmarks of a poor and ineffective missive.

It helps to jot down the points you need to make. Then set them out in logical order, so that one flows on to the next. Connect them with a general theme. Start with that theme; elaborate, point by point; then round off with the punch line or message. Explain what action should follow.

A good, sound ending to a well-constructed letter should bring business or whatever other results you seek. If the skeleton of the letter is sound, then even if the body is not as strong as it might be, the reader may not notice.

It was said of the great eighteenth-century lawyer, Sir William Blackstone: 'He it was that first gave the law the air of a science. He found it a skeleton and clothed it with life, colour and complexion; he embraced the cold statue and by his touch it grew into youth, health and beauty.'

You do not start with the letter of the law – you take the skeleton and you flesh it out. Create that foundation and clothe it with 'life, colour and complexion'.

2

Hail and farewell

A poor beginning or an inadequate end spoils any letter. Happily, the formalities, created by usage and etiquette, do much to ease the writer's path – if used correctly.

However much you may dislike the recipient of your letter, there is seldom any alternative to using 'Dear'. 'Hated Sir' is a splendid but (alas) unacceptable thought. To leave out the 'Dear' altogether is about as far as hostility (or formality) can decently go. Generally, such an opening denotes hostility implacable, indignation unquenchable. It has a Victorian ring and should be used only in very rare official and traditional notifications.

The use of a person's name without the opening endearment is generally only suitable for inter-office memos – or for deliberately offensive letters written (perhaps) to ex-partners or associates: 'Jenkins: I can no longer endure working with you . . .'

What matters is to make sure the name is right. If you do not take the trouble to find out the spelling of your correspondents' surnames, you can scarcely blame them if they feel that you are not really concerned with their business. Most people are inordinately sensitive about their names. Names represent the person. Not for nothing do political ruffians disfigure the names of their enemies on signs, posters and hoardings.

5

To make an error in a person's name is normally a sign of nothing but carelessness, which itself is hardly a good advertisement for you, your wares or your services. If in doubt, check. Phone the recipient's secretary. Is his name Philip with one 'l' or two? Is she 'Goldsmith' or 'Goldschmidt'? Is he 'Mr Ewan Richard' or 'Mr Evan Richards', 'Mr Harald Stewart' or 'Mr Harold Stuart'? Is his wife 'Caroline' or 'Carolyn'?

When writing to a woman, try to address her as she would wish – as 'Mrs' or 'Miss' or 'Ms'. Some women see no reason why their marital status should be revealed in their form of address, any more than it is for a man. 'Ms' is fine.

Of course, 'Dear Madam' (with an 'e' at the end only when writing to a French woman or to the wife of an ambassador or possibly to a brothel proprietress) may solve the problem, in the same way as 'Dear Sir' or 'Dear Sirs'. It is a very formal approach, and should generally be reserved for strictly formal letters, including those written to public bodies as such, as opposed to specific individuals within them. If you wish to deal with a person on a personal basis, then use his or her name.

Whether or not to address people by their first (or 'given') names is usually a matter of tact, dependent on your acquaintance or friendship with them. It is often a mistake to be too chummy, but if you can manage to be on first-name terms with business contacts then (in most cases) it will help to lower the fences between you. There are other occasions when keeping your distance may be an advantage. Judge each occasion, problem and recipient for yourself.

What then of the word 'my': 'My dear James' or 'My dear Mrs Brown'?

Use this approach with care. An upper-crust indication of warmth, it is sometimes taken by others to imply condescension. Customs differ, but if in doubt leave off the 'my'.

When you have addressed the recipient by his or her own name (surname or first name, as the case may be), you should almost invariably finish with 'Yours sincerely'. 'Yours very sincerely' is in order, if a little flowery. 'Yours ever' should be reserved for friends. 'Yours truly' is a handy variation used where the recipient is not well known to the writer.

'Yours faithfully' is the appropriate ending to a letter which begins 'Dear Sir'. It has replaced the old-fashioned 'I remain, your obedient servant', even in military letters.

The order of the final words is sometimes reversed, but this is usually an affectation. 'Faithfully yours' or 'Very sincerely yours' are rarities. Between friends 'Yours' on its own is common. 'As ever' or 'Yours ever' do good service. 'Sincerely' adds a touch of goodwill. But all these are variations on the basic themes of 'Dear Sir ... Yours faithfully' and 'Dear Mrs Brown ... Yours sincerely'.

Stick to the formalities and you are likely to start and finish correctly. You can then concentrate on the body of the letter.

If the beginning of your letter is weak, then the recipient will probably never reach its end. As with a speech, a telephone call or most other presentations, it's good to start with a warm-up. A few words of concern, friendship or introduction, but then the summary – your message.

Look at any first-class sales letter. The first sentence grips you and excites your interest. Poor openings are the direct route to the bin.

As Peter Ustinov said about after dinner speeches: 'You want a first-class opening; a first-class ending; and as little in between as possible!' He might have added that, subject to warm-ups or fond farewells, the start and the finish should be the same, even if the words alter. They are your message.

So pay attention to the hail and the farewell. The body of the letter will fall much more easily into shape.

3

Signatures and postscripts

There are people who make their living by interpreting characters as expressed in signatures. Loops, whirls and lines, we are told, all have meanings. Slope the words downwards and you are likely to be depressed and pessimistic; slope them up and the graphologist recognizes an extroverted optimist. Whether this has scientific foundation is both arguable and often argued. But the importance of a sensible signature is indisputable.

Your signature is your symbol. How do you make it? The height of modesty is to use an initial only: 'N. Smith'. Next in the scale of self-assertion, the forename and surname: 'Norman Smith'. Then comes the extra initial: 'Norman H. Smith'. Finally and most flourishing: 'Norman Halliwell Smith'.

Which signature will most impress depends on the nature of the letter. Suppose, for instance, the writer wants a job as a sales executive. The more extroverted he or she manages to appear, the more likely they are to be successful. Conversely, if a director is seeking a self-effacing assistant, they should look for someone with the quiet initials.

If you have to sign dozens of letters a day, you should develop a characteristic scrawl, to give your imprimatur with the minimum expenditure of time and effort. But a self-assured and swift symbol which impresses your staff or your regular correspondents may be counter-productive if the recipient expects to be treated with careful, considered respect. If you received an application for an executive position signed with a swirl and a squiggle, would you be impressed? Hardly.

One prominent executive says: 'A commercial mind should be like a pharmacy: a place for everything and everything in its place,' or, 'A tidy mind in a tidy body – that is the recipe for success.' If a signature is slovenly, then (rightly or wrongly) its reader may draw the obvious conclusion about the signatory.

Traders and merchants once took great pride in their seals. If they needed to sign, they pressed their ring or sealing rod or cylinder on to the wax and made their mark. Today, alas, the handsome, chiselled seal is gone. The only exception I know are prime ministers of India.

Your mark is your signature. It puts the final touch to your letters. So ensure that your touch is firm, appropriate and impressive. Many a good letter brings bad results because the writer is too lazy or thoughtless to add a fine, smart signature, suited to the nature and design of the words it rounds off.

Before you sign your letter, make sure that it includes everything that you have in mind. A PS may help, but it may also make afterthoughts too obvious. In sales letters, though, the first sentence and the PS are far the most likely to be read. Make sure that they have maximum impact.

You may correct omissions either by rewriting the letter, or by PS. Transpose anything of key importance to the body of the letter. And do spare time to read your letters, quietly and carefully, before you sign them.

There is nothing wrong with asking your secretary to put at the foot: 'Dictated by Mr James Jones but signed in his absence,' or simply: 'Diane Williams, pp James Jones'. Provided your correspondent will not be offended, all is well. But do make certain that the words for which you will be held responsible are accurate. You might even say: 'I hope that you will not mind this letter being signed in my absence by my secretary, so as to save a day – I am away tomorrow.'

If this happens often, then why not invest in a battery of those excellent stamps which give your secretary the choice between the different forms of 'pp' signatures. But be warned:

do not, repeat *not*, allow your secretary or anyone else to sign a letter on your behalf without your approval – other than in the rarest cases and when you have the most reliable pro-signatory.

Robert Browning, praising the glories of England in April, extolled 'the wise thrush; he sings each song twice over'. Wise letterwriters read the contents at least once over, with care. Far too many letters are sent unread, with a hastily scrawled signature.

4

Lists and schedules

A letter is a vehicle for ideas. Dry facts are better in an enclosure or appendix.

In general a letter is meant to be read. If the recipient is too bored or irritated to read it to the end – because it has been carelessly drafted, poorly written, badly word-processed, or is simply inept, inadequate or unfit for its purpose – then it has failed in its objective.

Lists or schedules are (usually, at least) intended to be consulted, perused, dipped into or extracted from, but unless the accompanying letter is apt, the list or schedule will be useless. Letters sell their writers' ideas, and often their goods or services. If you use schedules or enclosures, treat them too with respect. If you want them to bear good fruit, fertilize them with your time, care and consideration. You have no time to spare? Then make sure that someone else does so for you. Your enclosures should neither be, nor appear, as afterthoughts.

Some lists fit well into the body of the letter. If you are setting out a string of thoughts, then number them. Thus:

> Our reasons are:
> 1   the product is too new

> 2   the area is too limited
> 3   ...

Alternatively:

We are sure that you will appreciate the reductions in price in the following products:

> 1   ...
> and so on.

Lists can make a letter live. If you pack your facts or arguments all together into a paragraph you may succeed only in confusing both the opposition and yourself. List your thoughts, and if on reading them through you notice that logic is missing then you will know that you must redraft your letter, recast your list or change your mind accordingly.

Do remember to refer to the enclosure not only in the letter itself but also at its foot. The word 'enclosure' (or 'enc.') should do the trick. Or staple the letter to the accompanying documents so that they cannot go adrift. For every letter that is lost, there must be ten missing enclosures – which means a loss to the writer and irritation for the recipient. Make sure that both letter and enclosure are sent, and that they stick together.

In *How to Write Short Stories*, Ring Lardner advised: 'Many young writers make the mistake of enclosing a stamped, self-addressed envelope, big enough for the manuscript to come back in. This is too much temptation for the editor.'

Use stamped, addressed envelopes, and all other enclosures with discretion. Are they what *you* want? Are they really necessary? And are they there?

5

References on letters

References are as important for letters as they are for people. Careful, sensible, thoughtful referencing saves work, time and money. It informs the recipient immediately of the subject of your letter. It avoids mistakes and shows that you are a courteous correspondent. It also helps your own filing system to operate efficiently.

Every business letter should bear a general reference, after the opening or at the top of the page.

> Dear Ms Brown,
> Your order no. 25873

Or:

> Dear Mr Green,
> Directors' Loan Account

If you initiate the correspondence, you can generally choose the main reference. Try to make it specific. Where possible, for instance, refer to an order or contract by a number, and to people by their full name rather than by surname alone. The more precise your reference, the greater its usefulness.

15

If the correspondence is started by another party you will have their reference which you should repeat on your reply. But do not hesitate to add a reference of your own.

As well as the subject matter, the general reference should also indicate the identity of both the writer and the secretary or typist. There are two ways of doing this. It is usual to include the initials of each person. But if you would rather remain anonymous, you can allocate numbers or letters to every executive, partner and/or member of staff. If anyone wants to find out the identity of someone whose initials are on the notepaper, they can telephone the business number and ask. Where the letter is deliberately anonymous (as in some debt collecting missives) it may be preferable to include a reference which is clear to you but meaningless to your correspondent.

Obviously, copies of all important correspondence will be kept and filed (Chapter 44). The filing may be done by a highly trained secretary, which would probably be a waste of good and expensive time. More likely you will employ a filing clerk, or put office juniors on to the filing chore as part of their daily routine. You must then recognize the limitations of your employees, and allow for them. If you want copies to be readily available, they must appear on the right file.

Your referencing must be precise, clear and individual. Why not spare a few minutes of careful thought now, to avoid wasting time and energy in the future?

6

The layout

Well-marshalled thoughts expressed in well-chosen words may still convey the wrong impression if the letter is carelessly laid out. You should instruct your secretary and typists in the layout you require.

You may decide on some special house style. Eccentricity may be in your line. There are millionaires who slouch around in baggy trousers, multi-coloured neckties and without a penny in their pockets. I treasure letters from a great foreign statesman containing torrents of thought, poured out on the scruffiest of paper and with no paragraphs. The mighty are entitled to their foibles.

For the average successful executive – and for anyone who aspires to even higher levels in the business world – more prosaic behaviour is advisable. If you wish to reach or to stay at the top of the heap, appearances matter.

Executives generally realize the importance of being well groomed. They may even appreciate that their stationery should be as well turned out as they are. It is remarkable, though, how many people at the top take too little care about the layout of their letters.

Eccentricities apart, we recommend a standard style. Across

the top of the letter goes your heading. On one side (it matters not which) comes your address – or it may go across the top of the page. Slightly lower down, at the top and preferably on the opposite side to your address, put the recipient's address. The date goes where it balances best.

Then, 'Dear Bill' (or as the case may be) goes hard against the left margin. Next comes the reference, either in the centre or above. Margins on each side should be big enough to create a sense of space and to allow for the recipient's notes.

Lines should be double-spaced, with treble-spacing between paragraphs – indent sub-paragraphs. Finally the sign-off comes towards the middle of the page. Leave enough space for the signature. Underneath comes the name of the writer, and underneath that his or her job title or position.

If there are enclosures, add 'enclosure' (or the abbreviation 'enc.') in a bottom corner (see Chapter 4). If your letter will not fit on one page, use a continuation sheet rather than the back of the paper. All stationery should be ordered with continuation sheets of identical paper in the same size.

Here is a simple suggested layout:

The Jamestown plc
24 High Street, Jamesville, Beds MK41 3BT

Ref: GJ23   4 August 1998

Roger Brown & Co Ltd
38 Upper Street
Millbury
Wilts BA12 6PX

Dear Sirs

We thank you for your letter of 28 July. We shall be pleased to supply the goods required, if you will kindly provide the following further information:

      (a)
      (b)
      (c)

We look forward to an early reply and will despatch the goods immediately we receive your firm and detailed instructions.

Yours sincerely

Director

7

Acknowledgments and standard forms

The surest way to lose business and influence is not to reply to letters. The best way to reply to most is by adapting standard drafts.

Part of the art of ensuring prompt replies lies in an efficient office system. Are you satisfied that yours is as good as you can make it? Is incoming mail stamped with a date and shipped off smartly to the appropriate department? Do letters requiring an immediate answer always receive one? If the appropriate acknowledgment is inevitable, do you send it?

For instance: 'We thank you for your letter of ... which is receiving attention' or 'Thank you for your order. We shall deal with this as soon as possible' or 'Your communication is acknowledged. A reply will be sent shortly' or 'Thank you for your letter of ... This has been passed to ... for his attention.'

You need a batch of assorted acknowledgments, to be sent out as required. The person who sorts the mail should be able to decide which acknowledgment would be best, and should then put each letter in the appropriate tray, or say to a secretary, 'Form 1 for this, please ... Form 2 for that ...'

Standard forms there must be. These can be simple, brief and even on postcards. Or you may need something lengthier. For example, you may want to set out a series of possible common answers to a communication, all but one or two of which can be crossed out, for example:

> We regret that we have not received the order to which you refer.
>
> Please send a copy so that the matter may be dealt with as quickly as possible.
>
> We thank you for your order, but you do not give an address for delivery. Please do so, and we will despatch the goods within ... days/weeks/months.
>
> We regret that it is not possible to despatch orders overseas.
>
> We regret that the lines which you have ordered are for export only. We are sorry that we cannot be of assistance on this occasion.
>
> We have sent the goods you ordered by post/air/sea. They should reach you by about ... If you do not receive them by then, please contact us.
>
> We have checked our records. These indicate that the goods were despatched to you on ... We regret that they appear to have been lost in the post. We will send replacements without extra charge/investigate the matter/contact our carriers forthwith.
>
> Unfortunately, we are not able to open accounts for small amounts. If you will kindly send a cheque or postal order, we shall despatch at once.

You may add a standard PS: 'We have pleasure in enclosing our latest catalogue, which we hope will be of interest to you' or 'In case our other products may be of interest to you, we shall shortly be sending an up-to-date catalogue and price list' or 'It has been a pleasure to do business with you. We shall send you further details of new lines.'

It is, of course, important than even pro forma letters are well laid out (Chapter 6), and properly produced. Most offices these days have access to word processing or computer facilities, so it isn't difficult to achieve a professional appearance. And of course standard letters can be adapted easily to fit the particular circumstances.

There is a tremendous and growing variety of magnificent office machinery on the market, tailored to suit almost every commercial pocket (details in Appendix 2). Before you buy or rent make sure that servicing and spares are readily available.

Keep your eyes on the journals aimed at potential buyers (and often sent on a controlled circulation basis). If you are not on key mailing lists, ask your secretary to make a few phone calls so that you can be included. For details of current publications check *Benn's Media Directory* in every comprehensive reference library. Or ask business friends to pass over the journals they get.

Of course, the nature and number of the forms you should have in stock will vary according to your business. It may be helpful to prepare a variety of forms, to be adapted. Your word processor will cope.

Organize your letterwriting system carefully. Work out which letters can be reproduced, by whatever method is most appropriate. Set on one side those which are standard enough for draft precedents, and use them to lean on. If you do not want to invest in a collection of other people's drafts, then prepare one of your own. And leave for individual drafting only those letters which need special thought and attention.

8

Titles

The most common title (apart from a military or police rank) is
'Sir'. This is for knights and baronets of various orders and
degrees. In each case, unless you are on first-name terms with
them, the correct form of address is: 'Dear Sir Arthur', 'Dear Sir
Barnett' or as the case may be. But never 'Dear Sir Jones'.

To the wife of a knight or a Baronet you write 'Dear Jane' if
you know her well enough, or otherwise 'Dear Lady Jones'. The
same applies if she is the wife of Lord Jones.

If the lady has a title in her own right, then you also address
her as 'Dear Lady Jones' or (if she is a life peeress) 'Dear
Baroness Jones'. Or if you are on friendly but not first-name
terms, 'My dear Baroness' or 'Dear Baroness'.

The exception: the daughter of an earl, marquis or duke. She
has the courtesy title 'Lady Jane Jones' – so you write to her as
'Dear Lady Jane'.

The children of earls, viscounts and barons (life or heredi-
tary) have the courtesy title of 'The Honourable'. But you still
write to them as 'Dear Mr . . .' or 'Dear John' and never as 'Dear
Honourable Jones'. As a son of a peer I received many letters,
often computerized, starting (in Japanese fashion) 'Dear
Honourable Greville'.

Address for envelopes and letter tops: Hon. J. Green – or: The Hon. J. Green.

The woman with the female equivalent of a knighthood is a dame. She is not addressed as 'Lady' but as 'Dame' and, as with a knight, the title is combined with the first name and not her surname. Hence 'Dear Dame Jane' and *not* 'Dear Dame Jones'.

The highest rank before the peerage is baronet. On the envelope he is called: Sir James Jones, Bt, – but the 'Bt' is never used when addressing the gentleman. Equally, the lowest rank of the peerage is a baron. While, as we have seen, baronesses are often addressed by their title, the same does not apply to barons. They are always called 'Lord'. Hence: 'Dear Lord Jones'.

Higher up the lordly scale you may write 'Dear Viscount Jones' or 'Dear Lord Jones' – either would do. The same applies to earls, marquises and dukes.

The wife of a duke is a duchess; of a marquis, a marchioness; of an earl, a countess; of a viscount, a viscountess. Write to them as either: 'Dear Lady Jones' or 'Dear Countess Jones' – but just as you write to a duke as 'Your Grace', the same dignity is accorded to his wife. Never write 'Dear Duchess Jones'.

What, then, of priests and ministers of religion? 'Dear Mr Green'; 'Dear Father Green'; 'Dear Monsignor Green'; 'Dear Archbishop'.

The formal sign-offs should nowadays be simply 'Yours faithfully', or 'Yours truly' or 'Yours sincerely'.

Archbishops are generally addressed as 'His Grace the Lord Archbishop of ...' Address a bishop as 'The Right Reverend the Bishop of ...' or 'The Lord Bishop of ...', and start your letter 'My Lord Bishop' or 'Your Lordship'. Or if you know him, but not well enough for first names, 'Dear Bishop'.

Catholic dignitaries generally receive the same courtesies as their Protestant colleagues. When writing to a Cardinal, address the envelope 'His Eminence Cardinal ...', and start the letter with 'Your Eminence' or 'My Lord Cardinal'. Address a Catholic bishop as 'His Lordship the Bishop of ...' and start 'My Lord'. Address a Catholic priest as 'The Reverend Father ...' and start the letter 'Dear Father ...'.

Address The Chief Rabbi as 'The Very Reverend the Chief Rabbi'. Start your letter: 'Dear Chief Rabbi'. Otherwise, address rabbis as 'Rabbi Michael Cohen', or 'Rabbi Dr Michael Cohen' (as the case may be) and 'Dear Rabbi Cohen'. A Jewish minister who does not hold a rabbinical diploma is normally 'The Reverend Michael Cohen'. Start 'Dear Reverend Cohen' or 'Dear Mr Cohen'.

Now the law. If you know a judge well use his or her first name. If you know him or her reasonably well (but you are not on first-name terms), address him or her as 'Sir William' or 'Lady Jane' or 'Dame' or 'Judge Jones' (according to rank).

Writing to the Lord Chancellor? Address the envelope to 'The Rt. Hon. The Lord Chancellor' and start 'My Lord' or 'Dear Lord Chancellor'. The same style is appropriate for Lords of Appeal in Ordinary and the Lord Chief Justice.

Address the Master of the Rolls as 'The Rt. Hon. Lord . . .' 'The Rt. Hon. Sir . . . Master of the Rolls', or 'His Honour, The Master of the Rolls'.

Lord Justices of Appeal? Address them as 'The Rt. Hon. the Lord Justice . . .' or 'The Rt. Hon. Sir William . . . , Lord Justice of Appeal', and start 'Dear Sir' or 'Dear William'.

Lord Mayors? 'The Rt. Hon. The Lord Mayor of . . .', followed by 'My Lord'. An ordinary mayor is 'The Worshipful the Mayor of . . .', followed by 'Your Worship . . .'.

Address a Member of Parliament as 'Hilary Smith, MP'. Start, 'Dear Mr (or Mrs or Ms) Smith'.

Doctors of medicine should be addressed as 'Dr Roger Smith', except surgeons, who are 'Mr'. Hence: 'Roger Smith, FRCS . . . Dear Mr Smith . . .'.

Address a commissioned officer in one of the armed services by rank, together with decorations if any. To a Lieutenant-Colonel write 'Dear Colonel . . .' and never 'Dear Lt-Col . . .'. You may add the arm of service to the title of army officers and 'RN' to the address of naval officers.

Some say that when a letter is meant for more than one person, it should be addressed to only one of them. But 'Dear Mr and Mrs Jones' is much better. Leave out their degrees and the like. If you write to Sir James Smith, MP, alone, all is well, but omit the MP when writing to Sir James and Lady Smith. Or you could write to Sir James Smith MP and Lady Smith JP.

If you are addressing two or more people, other than husband and wife, use both their names: 'Mr Roger Brown and Ms Jane White', or 'Mrs Mary Green and Mrs Dorothy Red'.

What of 'Esquire' or 'Esq.'? Esquires were the rank below knights. The title was properly accorded to barristers and other presumed gentlemen, and by courtesy (or for the sake of flattery) to all men. It is now unnecessary and well on its way out.

And 'Master John Smith' for a child? Gone, along with my childhood: John Smith will do fine.

These are general rules. If in doubt, ask your secretary to telephone. 'How does the MD like to be addressed, please?' You will soon find the answer and avoid unnecessary irritation. People are touchy about their title and form of address. Courtesy and good results both require you to address your correspondent with all due care and in precisely the way they like best.

Part II

Style and Grammar

Part II

Style and Structure

9

Style

You judge applicants for a job by their clothes, their appearance and they way they speak or write – by body language and by their style. First impressions are vital, so style is all important. Once you know an applicant other factors become important – character and intellect, in particular. But at first, it is outward style that matters. If that is poor, then you may never discover that a gauche exterior conceals a mind of worth.

Equally, letterwriters should recognize the immediate impact of their correspondence. This means paying careful attention to stationery and printing (Chapter 42), to cleanness of type, to the skills of typists or secretaries and, above all, to the words themselves.

Grammar matters (see Chapter 10). If the writers are uneducated, unless they use and adapt precedents with loving care, the reader will know.

The best and most permanent answer is for the writer to become a reader. Do you enjoy reading? The first stage in correction is recognition of failings. If you can achieve this on your own, fine. But why is there such a prejudice among business people against formal education in letterwriting and literature generally?

Style cannot, of course, be detached from general layout. Letterwriters, like athletes, must make sure the start and finish are just right if they wish to win. The content must be clear, brief, lucid and pleasantly paragraphed (Chapters 12 and 14). Above all, the style of the words must reflect the intention of the letter.

What do you wear when you greet an executive from abroad? Dress must suit the occasion. The words of the letter-writer are the dress of thought. If they are inept, slovenly or ill-suited to the occasion, their style will destroy their impact. So fit style to circumstance.

Somerset Maugham said: 'A good style should show no sign of effort. What is written should seem a happy accident.' To put it in the words of Samuel Johnson (quoting a college tutor): 'Read over your compositions and wherever you meet a passage which you think is particularly fine, strike it out.'

10

Grammatical thoughts and ungrammatical words

Grammar matter – don't it?

'Between you and I,' the letter begins – and if the reader is a purist, there the correspondence ends. If the writer is applying for a job, prospects pall. Instead of achieving a confidential approach, the applicant has revealed ignorance of elementary English grammar.

Another common mistake: 'I am obliged to you for the courtesy extended to Mr Brown and I on our recent visit to your factory.' Courtesy extended to I?

'My staff and me are grateful . . .' So 'me' is grateful, is me? Someone does not understand the use of the accusative, and is guilty of a very common linguistic crime.

When you write letters, your accent disappears. Your speech may bear the marks of Belgravia or Bohemia, Brooklyn or Bermondsey, but your writing receives no overtones from your voice.

This may be an asset, for an accent is not always an advantage. But on paper everyone starts on the same level. Pens and word processors are classless instruments. The heavily accented words which you speak into your dictating machine emerge classless, sexless and without any indication of ethnic origin. At least, they should.

In practice, grammar is the giveaway. It places the seal of education on the letter of authority. Conversely, if writers express themselves ungrammatically, they stand revealed as uneducated. Not for nothing do we use the words 'unlettered' and 'illiterate' for those whose education is lacking.

We do not all have the same chances when we are young. Some top tycoons were forced into business at an early age. Once there, many read widely and improved themselves. Good companies buy good books if they want excellent executives.

How can you repair broken English? Here are three suggestions. First, read well and widely. Do you study financial columns and the form of both companies and horses? Fill in those long hours on train or bus or in the taxi by reading something more challenging. What about some history, or political commentary? Or perhaps the winner of this year's Booker Prize for fiction?

Schools too often spell 'literature' with a capital 'L'. They destroy young people's appetite for the classics by turning them into examination fodder. That's a pity. To write well, you should read widely. The greater your failings as a grammarian, the more you should marinate your mind in the rich wine of fine literature. Letterwriters who wish their words to have impact should read the letters of others. Many people have consigned their thoughts to letters. Some have caused or permitted their letters to be published. Read them.

Second, listen to good speeches. When you hear language well spoken, make a note. To copy the excellent is a mark of wisdom (provided only that you do not breach the laws of copyright – Chapter 48).

Finally, do not scoff at formal courses aimed at adults. If you do not want to attend a class, then have one custom-made. Find yourself a teacher. The money will be well spent. Or try a correspondence course: there are plenty about. Bear the cost or ask your employers to pay. You will benefit from learning and it is never too late to polish up your language.

11

Punctuation and self-expression

The object of most letters is to promote ideas. Whether you are buying or selling, hiring or firing, praising, decrying, apologizing or negotiating, you are expressing yourself on paper. As we shall see in Chapter 16, there are occasions when formality helps to keep the self hidden. Generally, though, say what you must (or what you wish) with clarity, and express *your* personality with *your* views.

We were taught at school that sentences must have verbs. Not necessarily. A sentence *usually* has a verb. Not always. Grammar has its place, but there is also room for the thrusting, vivid expression of informal thought.

Do not fear the lively phrase. Words may make a sentence with no verb. So may one word, on its own. Often. And effectively.

In general, sentences should not begin with 'And' or 'But'. But they often do. And to good effect. You are the writer. You make your rules. Yours is the meaning to be made clear, the personality to be expressed; your style is for you to choose.

Take punctuation. 'Period' is the pungent American name for the full stop. It indicates a break in thought. A comma marks a pause. A semicolon is half a colon and is useful for indicating the end of an item in a list. The colon comes between the semi-colon and the full stop: we have a pause in the flow of thought, but not for long.

The dash is a useful weapon – it indicates the break in a sentence (or list) longer than that signified by a series of dots ... These show that the thought has not ended, even though the sentence or the paragraph may have.

Careful punctuation breaks up a paragraph or a sentence. The breezier the style you choose, the more use you will make of the dot and the dash.

Suiting style to subject and to writer is more important even than the careful choice of words.

Modern usage often justifies the ignoring of old rules. Take the split infinitive, for example. I hate it, but that is a matter of taste. 'To carefully fix', 'to gently remind', 'to kindly honour', 'to swiftly reply', 'to eagerly await', 'to please reply': these are all quite common. But we all have our linguistic foibles, and I dislike the splitting of the infinitive for any purpose. My views on this may be old-fashioned. Before you split infinitives in your letters, though, bear in mind that the recipient of your letter – perhaps an important customer (actual or potential) or someone else whom you wish to please – may be as old-fashioned as I am.

In the margin of a state document, Winston Churchill once wrote, 'This is the sort of English up with which I will not put.' Me too.

12

Brevity

Those of us who are small – 'physically challenged' – are often reassured by kindly friends: 'The best things come in small packages ... A little person is a beautiful thing ... It's the size of the brain that counts ...' and so on.

For the man who craves those extra inches to dominate an audience or for the woman who regularly has to speak in public while resting her chin on the table, these thoughts give little consolation. But they do contain a germ of truth. Length is fine in its way, but it may be a nuisance. Tall people cannot stretch out in the bath or extend their legs in a sleeper or couchette. They can peer over the top of the crowd but seldom slide through it. As with people, so with letters.

There are times when a letter must be long to achieve its purpose. But generally, the shorter the words, the sentences and the letter, the better the results. Break even the longest epistle into brief sections. There is no excuse for the sentence that stretches into a paragraph, nor the paragraph that becomes a page.

Brevity is the soul of a good letter. Short, snappy, concise, clear and pungent paragraphs. Thoughts neatly packed into words with punch. Neat, lively expressions, shorn of padding and pomposity. These are the keys to successful correspondence.

The bore, the windbag, the person whom we would all go the longest distance to avoid, is also the writer whose letters we least like to read. 'Oh, him again,' you say, recognizing the tedious prose. 'I'll read it later ... if I have time.' The letter joins the ranks of the great unread.

In the world of journalism there are newspapers that pay by the word or the column inch. That puts a premium on padding. Many professional writers (like me) do their best to avoid this sort of yardstick. 'We only want 500 words' writes the editor. 'We pay £x per thousand words.' 'I shall be delighted to write your piece,' the journalist replies. 'But it will be harder for me to condense the material you want into 500 words than to produce a piece of 1000. I suggest that it would be fairer to pay the rate of £x + £y for the 500-word piece. It will take me longer to write and will cost more in care.' With luck, the editor will agree – professionals know that length and value are seldom the same. Quality counts. Brevity matters.

Churchill was once asked how long it took him to prepare a speech. 'If it's a two-hour speech,' he replied, 'ten minutes. If it's a ten-minute speech, two hours.'

In the world of public speaking there is classic advice: 'Stand up, speak up and then shut up.' But at least the spoken word is transitory. Unless you are on radio or television or in court, or you are a politician who produces some glorious gaffe – or, of course, unless you slander someone – your words will probably go unrecorded and unremembered. Business correspondents, though, have their words preserved in files, to be used in evidence if necessary. Keep those words short, accurate and to the point.

If you find that your letter is too long, take out your equivalent of the sub-editor's blue pencil. Peel away the extra words and leave them to stand on their own naked merits. If you are ashamed of them when they stand stripped, then think again. Redraft, rewrite, rethink. Excess verbiage offends, bores and muddles the reader and it also confounds the writer.

When General Eisenhower appointed Arthur Burns as Chairman of his Economic Advisors, Burns suggested sending the President a memo outlining plans to organize the flow of economic advice. Ike said, 'Keep it short. I can't read.' Burns replied, 'That's fine, Mr President. I can't write!' So they had a one-hour weekly conference instead.

A magazine once asked millionaire Paul Getty for a short article explaining his success. The editor enclosed his cheque for £200. The multi-millionaire wrote: 'Some people find oil. Others don't.'

Be brief, then. Or in the famous words of another oil man, 'If you don't strike oil quickly, stop boring!'

13

The choice of words

Businesses succeed or fail according to someone's choice of personnel. Two of the best ways of assessing people's calibre are by looking at their taste in friends and in books. Authors and speakers both make their impact through the words they use and the way they use them.

Consider, first, how one person's attitude may be described in many ways, some approving, some definitely not. Suppose that correspondents refuse to budge from a point of view. You may describe them as stubborn or stiff-necked, mulish, pigheaded, intractable, obdurate, hardline or intransigent, fanatical or thick. Clearly you disapprove.

You could equally well call them: dogged, pertinacious, determined, resolute, steady, constant, reliable: in other words, these people are courageous, and can be trusted. They will not bend before every blast, or trim their sails to even the strongest wind (splendid clichés those – Chapter 15).

So, adjectives must be handled with particular care. Meaning is fragile and may shatter if struck by the wrong expression.

But finding the right word (or the *mot juste* – the word which will do justice to the occasion) may be difficult. 'I can't put my finger on the expression I want. What was the word for that?

Oh, never mind,' says the letterwriter, using whatever expression comes to mind.

'What else was he to do?' you ask. Well, the writer could ask someone else to help pin down the word. Or consult any thesaurus. Here are tens of thousands of words, sorted into categories, with synonyms and antonyms, verbs and adjectives, nouns, pronouns – the lot. A thesaurus costs little and is available in hard cover or paperback. Every writer and speaker should keep one handy.

You also need a dictionary. It makes no difference whether 'you' are managing director or trainee manager. If letterwriting comes within your sphere, you need a collection of words and their meanings. The best is *The Shorter Oxford Dictionary* – a witty title, as you will discover when you try to lift its two enormous volumes!

For everyday use, try something really short – but people's abilities are judged by the extent of their vocabulary and the precision with which they use it. So if you do not know the meaning of a word, or are doubtful as to its exact meaning, check your dictionary.

Only a snob looks down on the illiterate, but lettered people (especially those who write letters for their living) invite derision if they mishandle words.

Those whose native tongue is not English have every excuse (Chapter 29 includes some horrible examples). But for those brought up in an English-speaking environment, it is essential that when their letters are funny it is by intention and not by mistake; that they should induce anger only by design; and that the words used should convey the intended meaning and have the desired impact.

To communicate accurately the meaning and intent of the writer – that's the object of a letter. Words are weapons. Select yours with as much care as possible.

14

Clarity

It is sad that the average letterwriter can generally spare so little time for the drafting of missives. Naturally, draftsmen and lawyers use precedents, but even these precedents must be adapted. Words which were adequate for one situation may need careful alteration for another.

Suppose you write to prospective employees telling them that you will provide them with houses. Are they then entitled to refuse flats or bungalows? You should have written 'residences' or 'living accommodation' or 'homes'.

Do your assistants have restraint clauses in their contracts of service (Chapter 55)? Do you describe with insufficient clarity the nature of the business which they are forbidden to follow when they leave you? Then the entire clause may be too vague to be enforceable.

You can probably find much better examples from your own disputes. Careful, precise use of the English language may take time in the short run but will eventually pay. Doubt destroys clarity; accurate wording eliminates doubt.

If you arrange for a builder to do work on your property but do not fix a price, a court would infer that you should pay a reasonable price. 'Obviously,' a judge would say, 'the parties must have intended that the contractors be paid for their efforts ... the inclusion of this clause was essential, to give the contract business efficacy.' In general, though, if the parties have not seen fit to provide for a situation, the courts will not do so for them. It is for you alone to make your deals.

Or suppose that you launch into freelance journalism. You agree to provide a feature for a trade paper. You should stipulate: 'If the article is returned to me within four weeks, then I shall accept it back without question. But if you keep it longer, it becomes yours.' This prevents any subsequent argument about the piece being sent 'on spec'.

The clearer your own thoughts, the better you will put them down on paper. The greater the clarity with which you write, the more likely it is that your writing will bring the results you seek. Unless you are deliberately trying to kick up dust so as to obscure some unpleasant issue, remember: unless your words are clear, they will not produce the effect you want.

15

Clichés

Good business demands modern marketing and merchandising. Poor, primitive or commonplace packaging or publicity spells death for the product. Good merchandise may be wasted on the market thanks to bad presentation. 'Trite,' you say. 'You preach to the converted.'

'Trite, maybe,' I answer. 'But you have yourself answered with a cliché – which proves my point.' The same business people who accept that merchandising material, packaging and presentation must be original, striking, vital and vivacious are still prepared to package their own thought in words so well chewed that they nauseate.

You should present your thoughts, like your products, in bright, original wording. Impossible? Then at least avoid the irritating cliché which slips so readily off the tongue but reveals lack of thought. 'The worse your case, the louder you should shout it,' proclaims the demagogue. Certainly the more unoriginal the thought, the more important is its disguise.

Have you ever listened to skilled politicians keeping their audience spellbound with words full of sound and fury but signifying nothing? Or sales people with a superlative spiel, wrapping their lines in clouds of persuasion to sell their goods hot, before their audience cools off?

How much more carefully, then, must writers mind their words and phrases. You want your correspondents to read your letters and to act on them? You wish them to provoke thought or sales or at least an appropriate and helpful reply? Then do the recipient the courtesy of avoiding the cliché.

Wit and humour help (Chapter 17). So do quotations. Ordinary, short, simple, Anglo-Saxon words will do as well (see Chapter 12).

Rule 1: Never use a long word where a short one will do. Verbosity impresses only the writer.

Rule 2: The more important your message, the fresher the words should appear.

Below are some of my most hated clichés.

> In this day and age.
> At this moment in time.
> This is an historic and unique occasion.
> This is a once only, unrepeatable and absolutely magnificent bargain offer.
> The tip of the iceberg.
> The buy of the year.
> We must give of our best.
> Each and every one of us must stand up and be counted.
> We are moving full speed ahead.
> Finally, and in conclusion.

And beware of 'Quite frankly ... frankly speaking ... to tell you the truth ... honestly ... I'll be frank with you ... genuinely ... sincerely speaking ...' They may be only turns of phrase, but they do suggest that the writer is well capable of not telling the truth. Really honest and frank people do not need to hang notices round their necks to tell the world of their integrity.

'We have the honour to be ... We beg to remain ... Your speedy response will oblige ...' Ugh!

Clichés, ancient or modern, are a menace. When you are next bored at a meeting, make your own list. 'Much water will flow under the bridge ... many bridges to be crossed ...' How time will fly by.

And take as a dreadful warning Churchill's description of Anthony Eden's speeches as consisting 'entirely of clichés – clichés old and new – everything from "God is love" to "Please adjust your dress before leaving".'

As Samuel Goldwyn so expertly misput it: 'For heaven's sake let's have some new clichés!'

16

Officialese and jargon

For years I corresponded with a company secretary. 'Dear Sir,' he invariably began. 'Yours faithfully,' he ended. Courteous but curt, and slightly overpowering in aloof dignity. Obviously, I thought, the writer is a tough, remote person with whom it would be a great mistake to tangle.

When I visited the company office for the first time I met the secretary – elderly, benign and gentle. He used his style of writing to disguise his personality and to keep his problems at bay. As he is an official, his letters reflect the policy of the board and the views of his superiors.

Official style, then, may provide a convenient shield; but official writers should still keep their words clear. There is no excuse for blunting meaning with useless jargon or clichés (Chapter 15).

Even if a letter is routine, it should still be crisply worded, with a clear message.

Some of the most horrible examples of officialese are in epistles from government departments or local authorities, others from bureaucrats of the business world.

'Vagueness may be necessary,' you say, 'to avoid making a decision, or to indicate that one has not been made. Clarity sometimes means unkindness while vagueness may cloak the unpleasant present with future hope.'

No one is saying that you must be tactless or cruel, but it is seldom kind to maroon the recipient of your letter in a sea of uncertainty, swelled by a flood of long, boring and mainly meaningless verbiage. Whether you write officially or unofficially, personally or impersonally, brevity and clarity should be the hallmarks of your style.

Use good precedents, where you can. Take your own drafts and rid them ruthlessly of every useless word, every ghastly jargon-ridden sentence. Shake off the padding and your bills for stationery, word processing or typing will shrink. A well-chosen word at the right time saves at least nine out of season.

Jargon may have its place as a cloak for the writer or even as shorthand in your trade or profession. But do not use it to muffle meaning, or to obscure the daylight from your thoughts.

17

Wit and humour

The humble pun has a bad name. Call it a play on words and it leaps back into fashion. Describe it as a *jeu de mots* and it is *à la mode*. Much of the best humour results from deft handling of the language. That is why jokes in foreign tongues are often so hard to enjoy.

Comedians or amusing speakers have many weapons at their disposal. Their art is part visual, part oral. The expression on their faces may matter as much as that which they give to their words. But writers depend on words alone. This may be an advantage. There are no distractions. Still ...

Novelists have the space to build up to comic situations. Even non-fiction writers have their humorous moments, using jokes or witticisms to point up morals or simply to enliven their style.

Letterwriters have less scope for humour and fewer words to play with. Their aim must be all the more sure. Puns have their place.

'You may bet your bottom dollar – if you will pardon the expression – that you will get more for your money with us.'

'This new line in pens will make its mark (in both senses of the phrase).'

49

'We were thinking of advertising these cut-price loos for customers who are none too flush – but perish the thought!'

Quick wit? Certainly not. A break, though, from the customary and boring. Even if the recipients groan at a pun, they may still wish they had perpetrated it themselves.

Irony and sarcasm

Sarcasm is supposed to be the lowest form of wit. Never mind. It has its moments. But it is designed to hurt: as there should never be hurt without design, use this form of attack with care.

'I know that you are immensely busy and incredibly overworked,' you write to a representative who is reputed to have done nothing of late save sit on his or her rear axle or on that of the company car. 'But do you think that you could possibly spare the time to sell some of our new lines? If so, we should be much obliged. If not, then I trust that you will not think it unreasonable of us if we terminate your appointment.' You have to be precise. 'Just one month more within which to reach your target . . .' If that does not stir him or her into extracting their digit, nothing will. It is sarcasm, and justified.

Irony is sarcasm with the sting removed. You place your tongue firmly in your cheek, smile gently and (with luck) produce a chuckle from your reader. For instance: 'No doubt there are many reasons for your poor sales record. When you say it could be worse, perhaps you are right. But we would require some proof of that proposition. Meanwhile, could you suggest how we can survive the summer?'

At the expense of others

'Taking the mickey' is a pastime to be enjoyed with care. Those who tease the most may enjoy it least when fun is poked at themselves.

Not all of us are 'good sports' when the joke is on us. There is a vast difference between the private laugh in a confidential note and the same witticism in a letter which may be seen by others. In the first case, the recipient may join in the laughter. The second may cause humiliation and upset.

Voluntary self-mockery is different from being locked into the stocks in the village green. So assess the occasion carefully before you provoke laughter which may turn into ill-humour, ill-will – and lost profits or lost friends.

Make jokes at your own expense. The cost is easier to bear.

The sting in the tail

A sting in the tail is the joy of all professional humorists. They lead you up one path, and then, when you think that you are in sight of the end, change direction so abruptly that you are left standing.

For speakers, as for tacticians, the element of surprise is vital. So is it also for writers of stories of suspense, detection or murder. It is a pity that it is not more often used by letter-writers seeking a dramatic and immediate effect. Some surprising examples:

> Over the course of the past five years, you have earned my friendship and appreciation by a long series of considerable kindnesses. Today presents a memorable occasion. I have the biggest favour yet to ask of you!

> I fully appreciate that the defects in the machinery which we sold to you led to some considerable unpleasantness between us, all of which was our fault – or, to be completely accurate, the fault of the machinery which we imported from France. If you tear up this letter, you would have every justification in doing so, particularly as I am about to ask a favour of you. Will you be kind enough to allow us to replace the defective machine, at absolutely no cost to yourselves, with another of a different manufacture, for which we are now the agents? Our good name and goodwill are vital to us and . . .

❖

I am writing to confirm that the time has come for your promised increase in salary. You have done a magnificent job and have earned the congratulations and appreciation of us all. The chairman, in particular, has asked me to say how greatly your services are valued.

However, you may wish to treat this letter as notice of termination or even of repudiation of your contract of service. As you know, the company is going through a period of very great financial crisis. So you and I are both being asked to accept a postponement of any increase in our salaries. I have accepted the situation. Will you? Please.

Would you come and see me? I have found this letter a very difficult one to write and would be immensely sorry to lose you as a colleague. As you know, we have great confidence in the future of the business. But the present is very difficult.

There are countless variations on this theme of the unexpected. Sometimes the twist is good for a laugh. Sometimes it is an introduction to an unpleasant shock. Sometimes there is only one gentle twist. Sometimes the letterwriter's path is full of hairpin bends. The reader's attention remains riveted to the writer's words. What's coming next?

Assuming that the writer is tactful, this off-beat approach will be appreciated. 'It's a miserable situation, but the man has a sense of humour . . .'; 'She is in a hell of a mess, isn't she? Still, things can't be too bad – at least she hasn't lost her sense of humour . . .'; 'At least life is never boring when you do business with him . . .' A tribute every time.

As always, judge the situation with care. Laughter and tears are close partners. A surprise that provokes a smile from one person may produce anger in another.

18

Similes and metaphors

Comparisons may be odious, but they provide a useful tool for the letterwriter. Use the word 'like' or 'as' and you have a simile.

Like:

> Our products are small and comparatively inexpensive but as necessary as the gears in a car.
> Like jewels in a royal crown ...
> As welcome as the rain that ends a drought.

Metaphors omit the 'like' or 'as'. For instance:

> We have struck gold.
> Bullseye!
> The demographic time-bomb.
> We fell at the first fence.
> This time we backed a winner.

The golden rule with metaphors is not to mix them. Mix your drinks and you become sick. Mix your metaphors and your letter becomes ridiculous:

> If you want to hedge your bets, then try reversing the batting order and your scheme may turn up trumps.

Business is a battlefield in which poor players are skittled out, often at the first fence.

I'm sorry that we cannot give you more credit, but the banks are at our throat and squeezing us until the pips come out.

It's an ill wind that blows no one any profits, when it comes to damping down demand by putting the handcuffs on embattled industry.

In a nutshell, good letters and mixed metaphors are as compatible as chalk and Chinese chopsticks, if you see what I mean.

Part III

Tact and Tactics

19

Modesty matters

Royalty and editors use the word 'we'. By adopting the first person, writers retain their personality and avoid the impersonal and soulless approach. By avoiding the singular they add strength to their purpose. Without saying so, they indicate that their views are shared by others: ... colleagues, the company, public opinion, the nation ...

There are two alternatives. The first is used by the French, the word '*on*' – 'one'. In English it sounds pompous and archaic. 'One does realize that, does one not?'

'One knows your difficulties, but one must take into account the current shortage of credit ...' One must indeed.

No, this will not do. 'One' is not pleased to accept your invitation to address the board – you are!

'We appreciate ...' would be fine if you are writing on behalf of the board. 'We have checked our note of the conversation with you ...' is plain silly if you are discussing your own conversation. 'We shall be pleased to come' is appropriate if you are bringing a partner or colleague, but if the acceptance is for yourself alone (assuming that you are not an extremely royal personage), the use of 'we' indicates either delusions of grandeur or a split personality, neither of which is likely to be appreciated by your hosts.

Some trades and professions have special usages. A one-person firm of solicitors may use the grand style, 'W ... & Co.', and the accountant may acknowledge: 'We are in receipt of your letter.'

Equally, if you are replying to a letter received on behalf of the company or firm, it is correct to answer: 'We thank you for your letter dated the 14th ...'. The thanks are those of the institution you represent and not yours personally. But if the letter came in an envelope addressed to you personally, and possibly marked 'Private and confidential' (for the legal rules, see Chapter 53), then you could almost certainly reply: 'I am grateful to you for writing to me ...', or 'I received your letter and will pass on your message to the board ...'

When you are addressed as an individual and written to on a personal level, stick to the singular. When you write on behalf of the business, the plural is generally more apt, more modest and implies greater strength behind your pen.

What, then, of the contents of the letter? How far in a letter should you avoid the vertical pronoun: I?

Public speakers must beware of the immodest, boring, self-satisfied impression conveyed by smug use of the first person singular. 'I came; I saw; I conquered', said Julius Caesar. But he was a dictator and could be excused. 'We came; we inspected; we took over', is more appropriate to the modern business world. If you came on your own and can honestly say that you in turn were conquered by the excellence of the hospitality received, then that would be different.

'I am very grateful to you for making my visit so interesting ... for sparing me so much of your own time ... for the trust you placed in me ...' You speak personally and sincerely. You write on your own behalf, to thank the recipient for kindness accorded to you.

You may also wish to poke fun at yourself. It is safer than ridiculing the foibles of others. The inimitable Lord Denning delighted in saying, 'When I used to sit and hear cases on my own, I could be sure that justice was done in my court. Now that I sit in the Court of Appeal, as one of a bench of three, the odds against justice being done are two to one.'

He could get away with it because he was a judge who combined extreme individuality and independence of mind with quiet and kindly courtesy. No one took his jest seriously – least of all himself.

Writers who obviously see themselves as sitting either upon the throne of God, or at His right hand, are fools. If you suffer from delusions of grandeur, then you would be wise to keep them to yourself or you may find yourself certified.

It is no excuse to protest, 'I am covering up for my inferiority complex.' Your correspondent may come to the same conclusion as the apocryphal psychiatrist – that you are just inferior!

It is not for you to shout aloud that you are great. One day you will descend, if only into retirement or the grave. Provoke jealousy by your words and you invite others to shake the tree – or even to pull it out by the roots. There is no need to go to the lengths of Uriah Heep: 'We are so very 'umble.' No one will believe you. Just write with humility, when you can. Modesty matters. Avoid that vertical pronoun and substitute the three-letter word 'you'. It is the most important word in the language of communication. Take care not to fall into the trap in which Edith Sitwell found herself: 'I have often wished that I had time to cultivate modesty ... but I am too busy thinking about myself.'

20
Flattery

It is stupid to emphasize 'the great I am' (Chapter 19), but the letterwriter should not hesitate to exploit recipients' inevitably high regard for themselves. 'To love oneself,' said Oscar Wilde, 'is the beginning of a lifelong romance.' Business people are as romantic as all others.

Note, first, that flattery is as useful when your fight is uphill or when you are faced with an unpleasant opponent, as it is when you are dealing with someone you admire. For instance:

I have admired your work for so long that I would hate to fall out with you now ...

You are a man most respected in this industry. So it genuinely grieves me to have to complain that ...

I fully appreciate your integrity and good intentions, but ...

I cannot believe that a man of your high standing would write as you have done, had he known the facts ...

I am sure that a company of the importance of yours would not willingly risk sacrificing its good name for the sake of ...

Do not underestimate your opposition. The essence of all sound flattery is apparent sincerity. People try to believe well of themselves, and if you share their admiration for their achievements you already have much in common. The higher and more powerful the executive, the more likely he or she is to be surrounded by adulators.

Still, where people achieve eminence in commerce you may expect them to be sound in judgement. Unless their power has corrupted them, they will be suspicious of the flatterer. Therefore use subtlety to succeed.

Overdo your flattery and it sounds perilously close to sarcasm. 'Of course you know best . . .'; 'You are as wise as the three kings . . .'; 'Naturally, I would not think of arguing with you . . .' Disastrous statements – and dangerous.

Here are some examples of nicely judged flattery, to use or adapt to business occasions.

> Experience has taught me to value your views. It is therefore with great diffidence that I suggest that on this occasion you are wrong . . .

> As you know, I rarely question your judgement – but have you considered . . .?

> I am extremely grateful for the trouble you took in writing to me. I know how very busy you are and your courtesy is immensely appreciated. I do apologize, therefore, for having to take issue with you on just one or two points.

> As you know, it is rare for me to take issue with you. I am new in this business and rely greatly upon the advice and guidance of good friends, such as yourself. But . . .

> I do appreciate your letter and I am sorry if I have caused offence where none was intended.

> I am sure that you will appreciate the difficulty of my position. I am not a free agent. I have to follow the instructions of my board . . .

> As usual, you are right. But dare I suggest . . .?

> It is very rarely that you are wrong. But with great hesitation and after much research, I have come to the conclusion this time that . . .

We are old friends and you know the high regard that I have for you. Our understanding over the years has been built on frankness and I hope that you will not misunderstand my motives nor regard this letter in any way as an attack on your judgement nor, still less, as a slur on your integrity . . .

You are a marvel! I followed your suggestion which achieved precisely the desired results. Thank you very much. Now for the future. Perhaps we can cooperate on . . .

21
The art of rudeness

A 'gentleman', or a 'lady', is a person who is never *unintentionally* rude. Intentional rudeness must be for some sensible purpose and designed for effect. In the theatre, it is taken for granted that an effect must be planned, created and perfected. In correspondence, planning equals preparation, precedent and thought. The effects which *require* the maximum of all three generally *receive* the minimum, because of anger or anxiety, hostility, dismay or urgency.

Attack usually produces that satisfying counterblast which relieves the writer's tension. You speak your mind so as to find a release for your anger. 'I must get it out of my system,' you say, or, 'Let's have it out ... I am not going to bottle it up ...'

The metaphors are apt. Just as the penitent relieves a guilty conscience by frankness in the confessional, and your friends feel so very much better after pouring out their troubles into your sympathetic ear, so hostility is released through expression. Take it out of your mind and you may release it from your memory. 'I'm a frank person,' you say. 'I don't believe in false pretences. Much better to have it out.' (There we go again.)

All this is psychologically and physiologically sound, but it spells potential commercial disaster. If you become overheated and are unable to apply a thoughtful, careful mind to unpleasant personal situations, then a moment's misery may turn into permanent disaster. Your rude letter may result in loss and litigation, annoyance and upset – an exercise that is profitless in every sense of the word.

In fact the first rule on writing rude letters may generally be summed up in one word – don't. If the situation requires straight talking, then arrange to meet the other person and talk. By the time of the meeting you will have cooled down and your anger may have turned to understanding, your wrath to forbearance, and your probable loss of friends and profits to potential future business. You will also have preserved your good name and your reputation as a person of dignity and restraint. At worst, if you do let fly, then at least your loss of self-possession will not be recorded on paper.

'Sorry,' you say. 'We have been patient for long enough. We keep receiving impertinent letters, and they require replies. I cannot allow rudeness to my staff without a vigorous response.' Or: 'I am sure that if I let off a real blast of indignation, I will bring them up short. Only rudeness will make them understand the seriousness of the position. A loud shout now may bring results.' Very well. Then sit back, consider and draft your words with care.

First, avoid foul language. You can be far more abusive with carefully considered, common words than with the four letter variety. So choose good words, with care. Remember the purpose for which they are intended and fire them with precise aim. Hone the razor's edge of your displeasure with the whetstone of your wit.

The turning of the cheek may be hard, but commercially it is generally sound. An angry word provokes a like response. Writers who give vent to their feelings on paper through lack of self-control may regret their words.

The right words in the right order stand a good chance of producing right results. But do use them with care. As Adlai Stevenson said, 'Man does not live by words alone, despite the fact that sometimes he has to eat them.'

If you *must* be rude, here are some examples for your consideration.

Rebuttal of persistent allegations

Yes, we have received your letters containing the same allegations. We replied to the first half-dozen, but no useful purpose would be served by our answering each one. Unlike the wine which you and I used to enjoy together before you saw fit to end our friendship, your allegations do not mellow or improve with age.

I have long restrained myself from expressing my thoughts. But may we now regard this unpleasant correspondence as at an end?

Riposte to rudeness

Having allowed several days to pass since receiving your impertinent letter, I still regard your sentiments as the most unpleasant that I have seen on paper for a very long time. In the circumstances, unless you see fit to acquire some unaccustomed humility and to apologize, this correspondence – together with our business and personal relationships – is permanently at an end.

Note The absence of a sign-off (Yours faithfully) is extremely offensive. Silence is sometimes a good deal more expressive and useful than words. Hence the best answer to many unpleasant letters is simply to put them on file, without reply.

Brevity – the soul of brusqueness

This correspondence is at an end.

Sarcasm – for a (former) friend

Will you oblige me and go to hell? When you arrive, you will certainly find yourself in congenial company. I wish you a safe journey.

Brief end to long friendship

Will we pay you more, you ask? No – a thousand times. But not again in writing. My regard for you is no longer worth even a postage stamp.

So we'll sue

Maybe when you come before a court you will learn courtesy. We propose to provide you with the opportunity as swiftly as possible. Your rudeness is intolerable. This correspondence will now cease. Instead, we have instructed our solicitors to issue proceedings against you forthwith.

Two-letter word

No.

Throwing their words back in their teeth

We append a schedule containing particulars of the speeches, impertinent exclamations and expressions of rudeness with which you have finally succeeded in destroying, utterly and per-manently, our regard for you and our business with you. If you were decent people we would expect a full apology. As it is, no doubt we shall hear nothing more from you. In one way, that will be a merciful relief.

Insensitivity incarnate

You are undoubtedly the closest approach to a rhinoceros ever to walk on two legs. If my poisoned darts have not yet penetrated your thick skin, then there is no hope for you. I can suggest that a holiday might help – and for all our sakes I would recommend a very long one, as far away from us as possible.

If you return to town in a recognizable human form, do please contact me. I shall be pleased to attempt to re-establish our relationship. You never know, we might be willing to place further orders with you. Do take care of yourself. There must be those who would miss you if you drove yourself to a premature grave.

Gloating over misfortune

Having received your letters – none of which, we thought, contained sufficient substance to merit a reply – we were not at all surprised to read of the judgement against you in the High Court last week. A few more like that and no doubt your company will be wound up. In that happy event, would you be kind enough to let us know, so that we may have the pleasure of attending the creditors' meeting?

Water off a sheep's back

As the PR people say, 'There's no substitute for wool.' If proof of this were required, your letter would serve admirably. It has no other apparent purpose.

The really effectively rude letters are usually those in which the words are barbed with wit or sarcasm, but which have an outward appearance of charm. The most devastating rudeness in speech is generally perpetrated with a sweet smile on the face.

To lose your temper is a sign of defeat. To lose it in court is often to lose your case. To lose it on paper is always to lose face and forfeit the full effect that your words could otherwise have.

Do keep cool, won't you? And address your letters with care, marking the envelopes (where appropriate) 'Personal and confidential'. Watch out for libel (Chapter 48), but rejoice that it is not defamatory to tell people to their faces precisely what you think of them. But when you dictate a letter, your secretary (at least) will know its contents. In this case there will have been 'publication', and if your words are defamatory you could be at risk.

22

Rude retorts

'How I wish I'd thought of that,' we all say, half an hour after we have made a lame reply to a rude remark. The devastating riposte, the unanswerable counter-thrust, the really rude retort salted with a touch of wit – these so seldom come to mind when you want them.

Replies to rudeness – in kind but with compound interest – are much easier when the initial attack was made by letter. So now suppose that your other cheek has been turned so often that it is sore, that you are determined to lash out at last. Then remember that there can be no general rule as to the best words for the occasion. You must match your wit to the words of your opponent. You must parry thrusts at the point of impact and hit back where it hurts most.

Here are a few sample suggestions:

> We suggest that the best way for you to appreciate the poverty of your case would be for you to read the words with which you have seen fit to clothe it. We are not referring to the grammatical or typing errors when we say that those words are as incomprehensible as they are discourteous.

> The emptiness of your threats is equalled only by the poverty of your product.

By resorting to blatant incivility you have not lowered the level of your case. It started below ground. May we respectfully suggest that you let it rest in peace? If you see fit to disinter it again, we shall have it cremated by our lawyers.

We are not at all surprised that you have descended to common abuse. A dried pea always rattles loudest.

How kind of you to reveal yourself so admirably on paper. Your charming words are appreciated for what they are – a smoke-screen.

We are, of course, sorely tempted to tell you to go straight to hell. But as this is an experience which you will doubtless have already had and enjoyed, no useful purpose would be served thereby. Instead, we recommend that you strive desperately for the other place. Were you to change your ways completely, you never know your luck. Meanwhile, we do not intend to allow you to make our business life into a purgatory.

If we were to descend to your level of abuse, our words would acquire the same odium as already attaches to yours. It is enough to say ...

If you manufactured fine sticks and stones, instead of cloth or ships, or shoes, or sealing wax, as the case may be, then your words might break our bones. As it is, we suggest that you re-read them. They provide a really excellent mirror, revealing the writer's mind ...

And your guide to replies:

1 Never descend to common abuse (it is too common to be effective).
2 Never lose your cool. In person, this may be difficult. On paper the heat of the moment can always be allowed to pass – there is no excuse for breaking this rule.
3 Bland discourtesy is best: the rude retort hidden in the sting of an apparently polite remark.
4 There is no better way to annoy your opponents than to laugh at them; apparently gentle words may have far greater effect than any vulgar obscenity.

There is, of course, one grave disadvantage to launching a written counter-attack. You will not be there when your letter is opened, so you will not be able to observe the effect it has on the enemy. Now, if you hear no more about the claim which your opponents were intending to make upon you, or if in contrast to your quiet, balanced and witty words, the crudity of their attacks becomes apparent – then you will probably have won.

23
Reprimands, warnings and complaints

The heat of the moment produces steam which should not be let off until you have cooled down. If you must explode, then do so out loud and resist the temptation to put your feelings into writing: the cliché 'least said, soonest mended' is true.

You may need to express your wrath on paper for the record – for instance, to prove why you rejected a product or dismissed an employee. Alternatively, you may prefer to write because you can choose your words with greater care, so that if they are used against you, you are less likely to regret them. Or you may wish the recipient to consider your views at leisure.

You must take particular care to match your words, which must always be calm, to the precise nature of the occasion (see also Chapter 24 on turning the other cheek and Chapter 22 on retorts to rudeness). Some useful examples:

> We enclose a schedule, setting out in detail the assurances you have given and broken in recent months. How much longer do you expect us to endure this sort of treatment?

> We have not completely given up hope of an improvement in our treatment at your hands, but we are about to do so.

Our patience is exhausted. This must be your last chance.

Please do not ignore this letter as you have seen fit to ignore those of ... Next time we shall not write, but take action.

Business correspondence is full of dreaded warnings of dire results. Some are gentle, some cruelly outspoken, some kind, some cruel. Here are some examples, to be threaded appropriately into your admonitory letter:

We have had to complain many times in the past concerning ... If we have cause to repeat our complaints in the future, then inevitably ...

We realize that this is the first complaint concerning your ... But you must appreciate that this is a matter of grave concern to us. We must warn you that if there is any repetition, then ...

I shall not warn you again ...

Please accept this warning in place of the action which we would have to take if there is any repetition in the future ...

The partnership between warnings and complaints is by now sadly obvious. Either may be fended off with the appropriate action, or often by an apology (Chapter 25). The art of complaining is itself worth careful study and is not always allied to a warning. Before you rattle your sabre, you must ensure that your adversary will be suitably impressed. If not, then threats and warnings are out, but you can still complain.

Complaints, then, may sometimes be allied with threats of warnings, but are often quite as effective on their own. Indeed, where a threat would be laughable or provocative it is in any event better left out.

The complainant's first task is to discover the best ear in which to pour his poison. Do you go straight to the top, knowing that as a result you will antagonize the junior person with whom you have dealt? Or do you exercise patience and keep your complaints on a lower level? Do you let off business steam to your trade association or Chamber of Commerce, or try your luck with your adversary's trade or professional association? Only experience can tell. Only after thought should you decide.

The complaint itself may be incensed or outraged in tone, or more in sorrow than in anger. All depends on the results you seek. Here are a few useful lines.

> We would be pleased to retain our association with you but ...
> We are sure that you personally could have no idea of ...
> We wish this were the first time that we had cause to complain about ...
> We are very anxious to avoid embroiling the board in ... but ...
> We are sorry to trouble you personally regarding the sins of your subordinates, but ...
> We have so far restrained ourselves from complaining, but ...
> We have complained many times. This is the last.

To be able to turn hostility into a friendship or a row into a firm, fat order – this is the mark of the skilful business letter-writer. Call it cynical if you like, but if through your calculated coolness you can make the other person feel bad, then you will avoid acquiring a needless enemy and you may exchange enmity for amity – with profit.

Turning the other cheek may require self-control, but if you cannot control yourself, you should not be in charge of others.

Even after the outbreak of hostilities, all is not necessarily lost. 'Softly, softly catchee monkee', says the oriental sage. Cool customers are admired, hotheads lose custom.

Therefore think carefully before you fire off any letter like those in the last chapters. Instead, re-read some of the flattering phrases in Chapter 20, and see if you cannot find the inspiration and the words to change your correspondent's wrath into gold.

24

In a tight corner

In times of trouble we often tend to draw analogies from the world of sport in general, and from fencing and boxing in particular: a cutting remark, a debating thrust, out for the count, hit below the belt, in a tight corner.

If you are in difficulties, you must choose your words with special care. In order to emerge unscathed, you have three possible courses of action. You can throw in the towel, trade blow for blow, or duck smartly under your opponent's fist and skip nimbly away.

For a change, let's take our first example from the speaker's world.* Suppose you are proposing a toast to the bride and groom. The bride's father is dead. The groom's parents are divorced. What do you do?

You can surrender by making no mention of the parents. This is abject cowardice, and would be generally regarded as such.

You can neatly duck the situation with a few carefully chosen sentences: 'The bride's father ... We wish he were here not only in spirit ... but he would have been proud and happy today ... How pleased we are that our groom's parents are both so well – and here together, celebrating with us ...'

*Details in *Janner's Complete Speechmaker* (5th edition)

Finally, you can take the bull by the horns (to take an analogy from another sport). Starting with the sort of comment given above, you can extend it into the appropriate elegy and eulogy. 'Let's face it, ladies and gentlemen – no occasion is completely perfect, no life without its problems. How sad we are that the bride's father is not here ... but we admire her mother doubly for the fortitude with which she bore her loss and especially for the courageous and splendid way in which she brought up the bride. The extent of her triumph is revealed by the radiance of our bride today.

'We know, too, that our groom's parents are united in their joy at his happiness and good fortune...'

Now suppose that you are writing to the family. You cannot attend their celebration. Normally, the less said about difficulties and differences, the better. 'With you in *our* happiness,' reads the cable from afar. 'How sad we are that we cannot join you,' goes the letter. But where the parties are close to you and you have to write at greater length, the above principles still apply.

> As you know, I was an old friend of your father. I know how delighted he would have been at your choice of bride.
>
> I am writing to you both, although I know that you are now apart – you will, I am sure, be together for the great day. I admire so much the way in which, despite your differences, you have always managed to be so understanding when it came to relationships with your son. You must be proud of him.

There are plenty of equivalent situations in business. The surrender is achieved by an apology (Chapter 25). The counter-attack is explained in Chapter 22 (on retorts to rudeness). The form of ducking away from trouble to be adopted will (as always) depend on the circumstances. Here are some useful opening gambits:

> We fully appreciate the circumstances which have led to the anger and disappointment expressed in your letter. There is another side to the story and we do hope that you will give it your consideration.
>
> You are right – but ...
>
> I do see your point of view – but am sure that you will give consideration to mine ...

Yes, we made a mistake – but in all good faith. The situation nevertheless remains that ...

We see your viewpoint. Now please do consider ours.

Your letter admirably sets out your case. It is only courteous, then, for us to set out as fully as possible the situation as we see it ...

Thank you so much for your promptness in dealing with our complaint. We appreciate your letter – and your viewpoint. We hope that, on reflection, you will agree that ...

No, we do not agree with you. But nevertheless ...

Those who use words as weapons employ very similar tactics to those of the fencer or boxer. You give way a little, so as to attack a lot. You retreat gently, so as to counter-attack with firmness. You at least pretend to see other people's viewpoints so that they will be prepared to consider yours. Alternatively, you politely disagree – and then show your magnanimity and/or good sense or goodwill by then offering a compromise on some point, however small.

The French put it well: '*Il faut reculer pour mieux sauter*' – you must withdraw, the better to leap forward.

There are occasions, of course, when you have your back to the wall, there is no room for retreat, all escape routes are cut off, you are up against the ropes ... Then remember the advice given to police officers: 'Tuck yourself neatly into the corner and use your fists, your knees, your truncheon ... At least if you are in that corner, they will not be able to get a knife in your back...' Unless, of course, they knock you unconscious and drag you out.

When desperate, try these gambits:

If you see fit to make these allegations to third parties, we shall have no hesitation in putting the matter into the hands of our solicitors.

Your threats are as empty as the premise upon which you base your allegations is groundless. Nevertheless, if you wish to take the matter further, we must refer you to our solicitors.

Your allegations are both impertinent and groundless. If they are repeated, we shall take such steps as our lawyers advise, to protect both our position and our good name.

If you are so ill-advised as to carry out your threats, then kindly direct all further correspondence to our solicitors.

In one, last, desperate attempt to remedy a situation which (we repeat) is not of our making, our Mr Jones will contact you and try to arrange some convenient time to visit your office.

Our chairman will be in touch with yours.

In the last resort, then, you have three options. First you can pass your correspondence to your lawyers, possibly in the hope that if you put on a sufficiently bold legal front your enemies will stay in their own trenches. You can cast aside pride or convention and try the personal approach, at whatever level seems best. Or you can remain silent.

An aged employee used to keep a little wooden sign hanging on the wall by her desk: 'Silence is golden,' it read. On occasion – indeed, on more occasions than most people realize – the adage is a good one. One way to emerge unscathed from a tight corner is to cover your face with your arms, crouch low, and pray for the sound of the bell.

25

Apologies

One of the most valuable words in the English language is – 'sorry'. If you are liable to lose your shirt, try displaying a white sheet. The effect on your adversaries can be quite startling. When people get what they want, when their pride is satisfied by the humbling of their opponent, they are willing to forgive a very great deal. For that reason every skilled letter-writer must know how to wield a dignified apology.

In the business world even apologies must be driven home with due care. If you have admitted fault or liability on paper – even by implication – you will be in trouble if you afterwards try to change your mind.

If you are involved in a road traffic accident, for instance, in which there is damage to person or property, an admission of liability may rob you of your insurance cover. Insurers do not wish to lose the chance to fight a case, if they see fit, through some premature genuflection by the insured.

Or take the common case of the supplier or contractor concerned to preserve the goodwill of a complaining customer.

The customer writes, moaning mightily. Instead of replying with a firm denial of liability and contradiction of the customer's allegations, the anxious supplier or contractor simply replies: 'I'm terribly sorry ... We greatly regret ... We shall do everything possible to put things right ...' The customer cannot be satisfied and refuses to pay the bill. The supplier or contractor sues.

'Look at the correspondence,' retorts the debtor. 'We put all our complaints into writing. They were never denied. On the contrary, the plaintiffs apologized and expressed their regrets. It's a weak excuse to say that they only did this to preserve our goodwill, isn't it?' Weak or not, it is commonly heard in courts.

So take care before you assuage the customer or client who takes umbrage, by apologizing in writing. If you feel that your best position is prostrate, arrange an interview, or apologize by telephone.* Any apology in writing is more likely to be used in evidence against you – unless you mark it 'without prejudice' (see Chapter 57).

That said, the art of graceful apology is still worth careful study. Here are some common and helpful forms of apology, to be incorporated into your letters when it is necessary or advisable to do penance on paper.

> Despite our every effort, these errors occurred – and we do apologize. We trust that no substantial or lasting harm was done and we are pleased to have the opportunity to put things right.

> We apologize most sincerely for any apparent discourtesy. None was intended.

> We greatly regret that you were offended by ... We are sorry that you took offence when none was intended.

> While we are extremely sorry that you felt that ... we must nevertheless point out that ...

> We are always anxious to have satisfied customers and therefore we are prepared to assist by ... We are in any event sorry that you were dissatisfied but we must point out that in making this offer, we do so without prejudice to our contention that ... and no liability whatsoever is admitted.

> I am asked by our board to say how much it is regretted that ... The Directors trust that you will accept their assurance that ...

> We apologize most sincerely for ...

---

*Be careful, though. Many conversations – especially by telephone – are recorded. See next chapter.

26

Lies – black and white

Some lies are forgivable. The letterwriter needs to know how best to disguise the truth. Tact is essential – in writing, as in speech.

Sometimes, the truth is just too horrible to tell. Imagine starting a letter: 'I finally decided to skip our lunch because I could not bear the thought of spending an hour in your company.' Or: 'The reason our chairman refused to speak to you is that he regards you as an unmitigated crook.' Much better to say: 'Terribly sorry, but I must ask you to be kind enough to postpone our lunch. The chairman has decided to descend upon us that very day.' Or: 'I do hope that you were not offended by any apparent discourtesy on X's part. Certainly none was intended.'

These of course, are examples of the white lie, designed by those without defence to avoid offence. When the truth would hurt or humiliate, even the moralist forgives the untruth. It's called tact. Here are the Ten Commandments of tact:

1 Check all previous correspondence and documents, to ensure that the truth is not apparent from your own previous writings.
2 Comb through your own recollection, and check with that

of your colleagues, to ensure that nothing has already been said to your correspondents which would now nail you as a truth bender.

3   If you can avoid putting the untruth on to paper, do so. The telephone is a useful instrument for the purpose, but remember: there may be a machine at the other end of the line recording your words. And while a contract made orally is as binding as any in writing, its existence and its terms are far harder to prove (see Chapter 54).

4   If you do decide to use the telephone as an instrument of untruth, then *you* have the conversation taped, if you can. One lie is bad enough, but to compound it with a contradictory untruth next time is unforgivable. A record of your words will prevent this.

5   If you must record your tactful untruth on paper, then take advantage of the ambiguities of the English language and try to make your words as vague as possible and capable of at least two interpretations.

6   Prepare an escape route in case of need. Maybe the transposition of a comma would alter the meaning back to the truth, so that you could then say: 'I do apologize for the misunderstanding, which was entirely due to a clerical error.'

7   If found out and there is no apparent excuse, be prepared to write: 'Although the letter went out under my reference, it was in fact written by my former assistant, Mr Jones, during my absence. I would like to emphasize the word "former". As a result of this episode, he has been dismissed.'

8   It is always better to have your doubtful letters signed by someone else, even in your name. Thus: 'You will observe that the signature at the foot is in my name but not in my writing. This was an inexcusable liberty taken by my then assistant, Mr Jones. He has been dismissed as a result. Thank you for drawing this matter to our attention.' (The fact that there never was a Mr Jones is irrelevant.)

9   When you receive a letter containing a lie, remember the above stratagems – they may be used against you; and study Chapter 27 for suitable ripostes.

10 Where the economy in truth is a large one, you have two alternatives: you may either build up to it by a series of minor fibs, or you can shout the big fib from the start in capital letters. Thus: 'I fear that you could not properly have read my previous letters to you, which state quite clearly that . . .' Or: 'I had not wanted to tell you, but it is necessary now to make a completely clean breast of it. The fact is that . . .' (The fact, of course, is fiction.)

Remember the eleventh commandment and keep it wholly. *Be not caught.*

You may, of course, take this chapter with as many pinches of salt as suits your literary palate. But like it or not, the 'tactful' lying letter is as much a fact of business and social life as the lying witness is a regular occupant of the witness-box.

To be 'economical with the truth' is not the preserve of distinguished civil servants, in witness boxes. Please note: I am *not* – repeat *not* – recommending this approach. Integrity is precious and lies destroy good names. Still, you should learn the tricks of their trade.

27

Replies to lies

Call someone a liar and you make an enemy. Suggest that they are mistaken and they may well agree. If you receive a letter containing untruths – however plainly stated – pause and look, before you leap to retaliate in kind. Self-restraint pays dividends.

There are many ways of turning aside untruths from others. A touch of sophistry is definitely justified. Churchill (who was forbidden by the rules of parliamentary debate to denounce a colleague as a liar) referred to an untruth as 'a terminological inexactitude'. A lie by any other name ... More recently the then Cabinet Secretary, now Lord Armstrong, admitted in court that he had been 'economical with the truth' (see previous chapter).

Here are some suggested ripostes to the lying letter:

> Whoever gave you the information upon which your letter is based are themselves in error. (*The height of tact, this – your correspondent saves face from the start and is given every opportunity to withdraw without humiliation – an essential in commercial warfare as in any other.*)

> I am sure that it is no fault of yours, but clearly your conclusions are founded on a misunderstanding of the facts.

I am most anxious that there should be no misunderstandings between us. If you would be kind enough to refer to our letter of ... you will see that the facts are not quite as you have stated.

I fear that our recollections of our conversation do not accord. I am quite sure that we agreed that ... I have confirmed this with our Mr Jones, who was present. *(What a pity that you did not confirm it in writing. Or maybe you did. Or if your letter is in response to one from him which purports to confirm the conversation but is wrong, then one up to you for reading his letter before you filed it. Beware of so-called letters of confirmation which are in fact travesties of the truth.)*

I am sorry, Jane, but you are wrong. We have known each other long enough to speak frankly, without ill-will or rancour. I am sure that the cause of the trouble was the report of ... I know that under no circumstances would you have written as you did, had you appreciated that ... I fear that Mr Green must have led you astray.

All my fault, I am sure – obviously I have not made myself clear.

As I am sure your letter was not intended to be offensive, I shall reply in full ...

I wish I had followed your suggestion and put everything into writing. As it is, your letter suggests that unless we can clear things up fairly fast, we shall become involved in misunderstandings which neither of us wants. Why don't we meet for lunch?

I am sorry, but I simply cannot accept the allegations contained in your letter. These are founded on an obvious misunderstanding of the situation. Perhaps it would help if I outlined our views, in full.

These gambits are handy whether or not your correspondent has in fact been rude or offensive to you. After all, he may have told an untruth or misstated the situation with a smile on his pen (to mix our metaphors – Chapter 18). But there are times when rudeness is too blatant to be ignored. This does not mean that you should reply in kind. Here are some suggested gentle retorts:

We have done business together for many years and I still hope that we shall do so in the future – to our mutual advantage. In the circumstances, I shall not reply to your letter in like tone.

We have known each other for a long time and I really am shocked at the way in which you have seen fit to write. I think that someone is stirring the pot. The statements you make are untrue, but I am sure that these have arisen out of a misunderstanding, probably created by someone who hopes to drive us apart. Let us keep cool. I would be happy to meet you.

Surely it does not help the situation to write as you have done? In the circumstances, I shall resist the temptation to answer in like terms. It is a great pity that we are at loggerheads. This is, I am convinced, unnecessary. Let's try to make a new start, shall we?

If I do not answer your rude letter with an even ruder one, I hope that you will not take this as a sign of weakness. Equally, it would not help if I were to indicate my views about the misstatements which turn your letter into fiction. I would prefer that we try to revert to where we were before this correspondence started, in the hope that we can clear up the mess without the entire matter landing in the hands of lawyers. If we are no longer to do business together, so be it. But at least let us be civilized. It may help if I set out the facts as I see them.

You may have noticed that in this sort of reply the cliché comes in handy. The set phrase helps conceal the unsettled temper. 'You are in error ... You are mistaken ... We regret that you are misinformed ...' – all much better than 'you are wrong', and almost invariably preferable to 'you are a damned liar' – even if that is the truth. As every merchandiser and PR person knows, both a product and its presentation need careful wrapping up. The noble art of the skilful wrap-up should form part of every course on business, with special reference to the reply to lies.

28

On a personal note

In general, personal notes should be written by hand. This is far more troublesome and time-consuming than dictating – which is one reason why the handwritten note is so appreciated. Whether you are sending congratulations or commiserations, apologies or thanks, three lines written are often better than three pages typed.

If you cannot spare the time to write (literally), then at least make sure that you 'top and tail'. Have the body of the letter word-processed, if you must, but write in 'Dear Joe' and 'Yours sincerely, Martin'. Or at least add a handwritten PS.

Naturally, it is helpful if your writing can be read. If you are taking time to write, then at least spare the extra few moments to do so without the apparent haste which the erratic line and the smudged sentence so clearly indicate.

If you want a good reason for careful writing (courtesy apart), then remember that there are those who specialize in reading character from script. This may, of course, be a good reason for having your letters typed, but it also explains why many prospective employers, when advertising for staff, say 'Please send full details in your own handwriting.'

The object of a personal note is to convey personal thoughts.

The more impersonal the form of the note typed (word-processed or, much worse, duplicated), the less its effect. Conversely, any personal touches are welcome.

The method of writing matters. So also does the method of delivery. If you can manage to send your missive by hand, marking the envelope accordingly, you are setting a special seal of importance, thought and urgency on the contents. Recipients also see that you have taken trouble.

Handwritten notes should generally be separate, however short they are, although there are times when they can be added to something else. If, for instance, you have to send a printed circular to a friend, then add a note at the foot in your own handwriting. Or perhaps invitations are going out for a company party. These have to be printed. Then add a few words at the bottom: 'Do come! – Johnny,' or 'I look forward to seeing you – Mary.'

Do you especially want the recipient to come to a meeting? Add a footnote: 'PLEASE be there – I need you!' or, 'A full turn-out is vital. I know you are terribly busy, but I would be immensely grateful if you could come.'

Impersonal notes are destined for the fire or waste-paper basket. The more personal your tone, the greater your impact. So be personal.

29

Letters overseas

The traditional caricature of the English? At home, polite but humourless, sitting silently in a train, talking to no one; abroad, a phlegmatic figure, waiting silently for a foreigner who speaks English. In fact, this person will soon have followed the dodo into extinction. Nowadays, most English abroad will 'have a go' at the foreign language, even at the risk of making fools of themselves. They know that if they wish to make friends or to influence business in their direction, they must make the necessary effort.

Take Americans entering world markets. They will arrange crash courses covering not only the foreign languages needed but also local customs. They know that customers have customs which you must respect.

Curiously, these rules, which are self-evident in courteous business conduct on the personal level, are often disregarded when people put pen to paper or mouth to dictating machine.

Peruse almost any file of correspondence with a foreigner. The letters will be filled with jargon scarcely comprehensible even to the writer's compatriots. Jargon is bad enough on the home market (where it may do grave damage – see Chapter 16), but it can leave overseas customers in despair.

You may, of course, point with justification to painfully correct and cliché-ridden letters received from Germany, for instance, with verbs placed at the end of lengthy sentences and the writer having, as always, the honour to be your most obedient and respectful servant. To every nation its style, and if you wish to do business with them you must respect that style.

What then is the answer? There are two possibilities only. Either write your letters in – or have them translated into – the foreign language concerned, or use the English language in the way that is least alarming to foreigners.

Unless you are yourself expert in the nuances of the foreign tongue, my advice is to keep it for convivial speech or tabletalk, and not for business letters (except, perhaps, for friendly, non-business postscripts). There are enough misunderstandings in international affairs without your adding to them by mangling the language on paper. By all means flatter and please foreigners by conversing in their own language, but let your efforts stop there. If you employ interpreters or translators make sure they are bilingual and totally trustworthy. Only those who are sensitive to shades of meaning in *both* languages can do the job properly. And to be on the safe side, send the English version together with the translation.

Truly bilingual people are rare. Some of the difficulties are obvious even when you consider the differences in idioms used by people who ostensibly all speak English. I once saw the end of an American romance when I assured my New York girlfriend that I would 'knock her up at six'. Differences between lifts and elevators, braces and suspenders, pavements and sidewalks – all should be known.

Or take pronunciation. The overseas lecturer who explained how invading forces landed on the beaches and created peace for the inhabitants caused chaos by his charming pronunciation of the words 'beaches' and 'peace'. Letterwriters may at least rejoice that foreign words on paper do not have to be pronounced. But spelling may be crucial.

When you do write in English, all the rules in Chapters 12 and 14 apply with even more force. The simpler the language the less likely it is that the recipient will misunderstand; the shorter the sentences and the clearer the thoughts, the better the business you are likely to do.

Like all rules, these have their exceptions. If you are trying to delay, then selective sentences of specially contrived jargon may send the foreign correspondent scuttling for a dictionary or (better still) off to join the queue at the local translators.

If you doubt these suggestions, consider how much pleasanter and more profitable your dealings with your overseas correspondents would be if they treated you as I advise you to treat them. Not for nothing did the Tower of Babel collapse. Some sensible goodwill, simple phrases, trained translators – these might have kept the building erect to this day.

30

The human races

The recipients of your letters may be as prejudiced as you are, but they may not be so ready to recognize their defects. Or you may be dealing with someone of genuinely open and intelligent mind. So if you are tempted to salt your letters with remarks which might be interpreted as racist, or otherwise offensive, don't.

Take the Scots or the Irish for instance. They may delight in making jokes about avarice and whisky or brogue and the Blarney Stone (as the case may be), but this does not mean that they will like it when the jokes come from others. The best raconteurs of Jewish stories are themselves Jewish. We Jews are entitled to laugh at our own miseries and we recognize that it is our sense of humour, as much as any other quality, that has enabled us to survive. But no Jew appreciates an anti-Semitic joke coming from a non-Jew, even a close friend. Jokes about popes, pills and priests may go down splendidly if told by Catholics. But in a letter, even a hint of anti-Catholic prejudice may spell ruin.

'We would never Welsh on you.' 'Like an Arab market place.' 'Eeny, meeny, miny, mo' – expunge them all.

Normally, no civilized person wishes to cause unnecessary

offence (for the correct use of rudeness, see Chapter 21). Avoid sexist jokes and references: they are more likely to offend than amuse (and don't forget, that letter you started 'Dear Sir' is quite likely to land on a woman executive's desk).

There is an ancient rabbinical saying: 'Respect goes before the law.' Respect for the ways, feelings, attitudes and ideas of others is even more important than adhering to the letter of the law.

So make jokes against yourself or about your race or your religion, and the world will laugh with you. But not against, or about, other people's.

31

Following up

The matador hopes to kill his quarry with the first thrust of his sword. If he succeeds, he may be awarded one or both ears and (most exceptionally) the tail as well. The letterwriter may make a killing at the first blow – so earning profit, an appointment, a prize or some other happy result. But a follow-up may be essential, to clinch, to close, to succeed.

The following is a list of memory-joggers, adaptable to most occasions.

> I did appreciate the time you gave to me. Could you spare, please, just a little more to answer my letters?

> At the risk of becoming a bore, may I please remind you ...

> I do know how very busy you are, so I would doubly value a reply to ...

> I do not seem to have received a reply to my letter of the ... I am quite sure that this is due to an oversight on your part, but your early attention would be greatly appreciated.

> I would not dream of pressing you if I were not myself under great pressure.

I fear that unless you can kindly make your decision shortly, I shall most reluctantly be forced to ...

I know that I have written to you before concerning ... I hope you will not mind my doing so once again.

I refer to my letters of the ..., the ... and the ... Could I now *please* have the courtesy of a reply?

I fear that unless I receive a reply to my letters within seven days, I shall have to ...

I am sure that you intend no discourtesy by ignoring my letters, but on reflection you will agree I hope that it will not help us.

You frame your reminder to suit its purpose. If your first effort fails, do not be afraid to try again. The more trying you become, the more likely you are to succeed.

Part IV

Letters for Occasions

32

Introductions and references

The greatest favour that you can do for most people is to provide them with the right sort of introduction. 'What matters to good advocates is not to know their law,' said the wise old practitioner, 'but to know their judges.' In every field it is *whom* more than *what* you know that matters. Match the introduction or reference to victim or recipient.

Letters of introduction and reference require special care. If they are inaccurate or negligent, you could be in trouble. If a reference is defamatory, then you could rely upon the defence of 'qualified privilege', but if 'malice' is alleged against you, it could be a long, hard fight. Nor is a disclaimer a defence against a negligence claim by your victim (see Chapter 52).

The best introductions and testimonials are the briefest. An excess of superlatives destroys the effect. I suggest:

To whom it may concern

I am pleased to recommend Mr James Blank, who has been employed as a ... in my department for ... years. I have found him reliable, diligent, cheerful and helpful. We are very sorry that he is leaving us.

I would be grateful for any assistance you could give to Mrs Mary Brown, who has supplied this company with . . . over the course of . . . years. We have now moved into a different line of production and can no longer make use of her products. However, we are happy to recommend her and them to you.

I am pleased to provide a reference for William Jones. He is a man of energy, tact and initiative who is leaving us because we have been obliged to close down his department. We cannot offer him the prospects of promotion which he deserves. If you require any further information, please do not hesitate to contact me personally.

Finally, remember to disclaim, loudly and clearly, in all appropriate cases. If a reference or testimonial is given carelessly and causes damage to the recipient, it will be no answer for you to say that you were not paid for providing it (see Chapter 48).
    Examples:

While we are pleased to assist by providing references/information/advice, these are given on the strict understanding that no legal liability of any sort is accepted in respect thereof, by the company, its servants or agents.

The above reference/testimonial is given without legal responsibility.

No responsibility can be accepted in respect of the above reference/testimonial.

My favourite reference is to be avoided.

I am pleased to recommend Roger Green for any other job!

33

Letters that sell

Whatever your business, you must sell. Whether you are a manufacturer, a wholesaler, a retailer, a stockist of services – whatever your trade, industry or profession – the moment you stop selling, your business starts dying.

Some sales are made face to face. The larger the item, the more profitable it becomes to devote personal time to seeing the potential customer. You may make direct sales through advertisements (not the province of this book) or by direct mail. But many of the best sales are won through the right letters landing on the client's or customer's desk, or are lost because those letters are off the mark.

There is no such special creature as the sales letter. Nearly every letter is trying to sell something. Maybe it is goods or services; or your case for an overdraft or an indulgence; or simply your good name or goodwill.

It follows that if you seek a direct or indirect sale through the mail, all the normal rules in this book apply to your letter. For instance, you must pay proper attention to stationery (Chapter 42) and to the envelope (Chapter 43); to the presentation of the contents and in particular to the methods of reproduction

(Appendix 2). Your message must be concise (Chapter 12) – 'Brevity is the soul of sales', as one top merchandiser puts it. You must consider ways of signing on and off (Chapter 2) and pay careful heed to opening and closing with a punch (Chapter 1). You should use flattery (Chapter 20), humour (Chapter 17) and sound and sensible modern grammar (Chapters 10 and 11).

There are, though, some special problems. After a careful study of your market and bearing in mind the limitations of letterwriting, what is your best approach to the sale of your particular goods or services?

If you have your prospects sitting across the lunch table, if you have 'found them at their desks', if you are chatting to them in your shop or showroom or at your stand at a trade fair, then if one approach fails you can try another. You can observe the effect that your words are having and adapt your style or approach accordingly. You can argue, cajole, pit your wits against those of your prospect. Naturally, you try to avoid the wrong initial approach. But because the prospect stares you (literally) in the face, you often have a second chance when the first proves unrewarding.

In that case, you sacrifice your own time in the hope of a direct sale. Unless the customer or the sale is important, the exercise is not worth your time. The beauty of a letter is that it can be prepared swiftly and cheaply. In return, you must watch your words with even greater care than if you were speaking. If your overture falls flat, there will be no encore.

Curiously, many top-selling organizations which work out a careful routine for their sales staff prepare little or none for their sales letters. Sales staff are taught what to say and how to say it, how to respond to each reaction of their customers, the precise approach to each stage of the selling process. They learn how to insert their feet in the doors of commerce and how to keep them there.

All this makes good sales sense. Why, then, do the same businesses become so sloppy when they take to the mails for sales? After all, it is much easier to provide drafts for letters than precedents for speeches. Reproduction on paper is far more certain than in either speech or procreation. If you are prepared to devote enough time and thought to your sales letters, you still will not be able to guarantee that they will 'click' every time, but they should generate far better results than you have at the moment.

What should you include in your sales literature? Is yours a case for the soft sell or do you plunge right in and hit your customer hard? Do you make a special introductory offer or is yours a 'once only' effort? Do you lay emphasis on price or quality, on past achievements or future prospects? Should you adopt a style that is formal or colloquial?

Precisely because of the infinite variety of circumstances, products and services, of sellers and buyers, of needs and impulses, no one can answer these questions for you. You must do so for yourself.

The only golden rule is to spend as much time thinking about your reader as you do about your own products or services. Many sales letters are self-regarding hymns of praise. They are so busy extolling the remarkable features of whatever it is they are trying to sell that the readers become invisible and their needs are ignored.

One of the principles of successful sales is to draw a clear distinction between *features* and *benefits*. In essence, a feature is what the product or service *is*; a benefit is what the product or service *can do* for the buyer.

Place yourself in the role of recipient. When you receive a mailing shot, why and when do you read it – and what makes you buy?

As with all letterwriting, the look of the envelope must be right. You must consider its colour and quality, the style of the address, and whether a computerized label will suffice.

If you are writing to a company, then whom should you try to reach? The managing director, company secretary or buyer, the works, personnel or shop manager – or who else? As with all selling, you must identify your market and zoom in on the individual who has the authority to place the order. If you can target the person by name then do so, but make sure that you get the name absolutely correct. Otherwise, keep to the title; 'Dear Colleague/Delegate/Buyer/Executive/Fellow Manager . . .'

You may be lucky and have your own list of potential buyers. Perhaps you have built it up from previous business. Remember, though: people change jobs within organizations and move to other companies. Your own lists need frequent updating.

The next problem is how to start your letter. If you are using a computer or word processor (see Appendix 2) you may start 'Dear Mr Brown' and then the machine can do the work for you. Otherwise you may address the shot to the managing director and start your letter 'Dear Sir or Madam'. Better: 'Dear MD' or 'CE'. Do 'top and tail' your letters individually whenever you can. But make sure the name is 100% accurate.

What should you put into the letter and what is better in an enclosure? The letter should be brief and the accompanying literature should contain the detail. The letter must draw the customer's attention to the enclosure – and, with luck, to the order form.

The letter itself should be brisk. As always, the first sentence is of vital importance. If that is wrong you pronounce the final sentence on the letter. Conversely, if the initial impact is right you will arouse the interest of your reader and perhaps be on the way to a sale.

The first sentence should encapsulate the purpose of the letter. Tell the reader what you are offering and why it is unique.

You could pay professional copywriters to prepare your letter, but there are few who do the job really well. Even if they have great skill with words, they will know neither the product nor the market as you do. The best way is to do the job yourself and save the cost of the experts. If you have to use them, at least prepare drafts, review their rewrites and always insist on seeing the final draft and, if it is to be printed, a proof.

End on a climax. Round off with your sales pitch. Sign. Then add a PS. Research shows that this has the next best impact to the first sentence.

Make sure that your reply card or order form is detachable; that any offer you make is easy to accept; that any reply is kept simple; and that the letter is accurate, well presented and professional. The sales letter must refer to the enclosures. The higher the aim the more dignified and prestigious the stationery and the wording should be.

Will the letter work? Try a test mailing and find out.

Which lists best suit your purpose? Probably those which you have yourself built up of old and satisfied customers and friends. Otherwise try local sources: the telephone directory; classified sections; trade directories; town guides; lists which chambers of commerce may provide, free or for a fee.

On the national level, use telephone and trade directories. *The Times 1 000* provides details of leading business enterprises, and the guide to *Key British Enterprises* selects some 50 000 of the same. *Kelly's* directory lists some 90 000 names.

If you are aiming at a specialist market you will find that most have at least one specialist directory. Professional, learned and other bodies, institutes and societies produce their own registers.

Other markets require other methods, for instance spotting from the electoral register young people who will be 18 that year, or checking engagements and marriages in the columns of the local paper.

You may also buy or rent lists from direct mail houses or other specialized organizations. Others will sell you their lists at a fee or even swop their list for yours. At the end of this chapter is a list of some of the organizations which may provide you with the help you need, from whom you may buy directories or lists, or from whom you can at least obtain guidance or quotes.

Remember, direct-mail selling is a highly skilled and competitive affair. Before you launch into it, work out the costings. How much will you have to spend and what return will make the outlay worthwhile? What are the current postal rates, how are they likely to change, and what are the maximum sizes and weights of mailing shots?

Would it be worth your while to arrange for sampling, testing or market research to be carried out before you make your investment? Letterwriting on a large scale involves massive potential expense and risk.

The same rules that apply to domestic direct mail apply with even more force to direct mail overseas, where the costs are higher.

Sales letters of any kind require thought and preparation. You should not 'bash out' a letter, 'tear off' a memorandum to a customer, 'dictate a quick "mailing shot" ' – and then have it completed and polished by some subordinate or maybe signed in your absence. These letters count. They must suit your purpose *and* meet the needs of your reader. If they don't, your letters are likely to end up 'deep-filed' at the bottom of your reader's waste-paper basket.

34

Applying for posts

As a preliminary to any legal arrangement, as an opener of the contractual door, as a means of gaining entrance to the board-room, the office, the factory or any other place of work – the letter of application is most important. Of all the business letters that you have to write, the job application is the hardest. You are selling yourself by direct mail, or at least using the mail to invite an offer for your services. So consider the best way to perform this immodest task.

What made you write? The answer provides a simple, brief, effective and invariable opening gambit:

> I am replying to your advertisement in today's *Gazette* ...

> I understand from our mutual friend, James Brown, that you have a vacancy for the position of ...

> I have been referred to you by ...

The chairman, managing director, company secretary, human resources director – the potential employers or their representative – will not at this stage toss your letter into the waste-paper basket. The door is unlatched. Apply your shoulder:

I am most interested in the possibilities which your position offers.

I appreciate the responsibilities which the successful applicant would bear and I find these exciting.

The challenge offered by your post is one that I would welcome.

Members of Parliament, of local authorities and of the board of directors of companies, all share a duty. They must declare any interest in matters under discussion. If, for instance, you have a financial interest in a contract which your board is considering, then you must say so. If you wish to win a place on that board and are applying from the outside, then the more interested you appear in the work that is offered, the more likely you are to be successful. Employers seek enthusiasm.

You have explained the origin of your application and your interest in the work. You have introduced the subject. Now introduce yourself.

I am 35 years of age, and have been employed in responsible positions in the industry, ever since I completed my Doctorate in ... at ... University.

I am at present a director of ... Ltd. You will appreciate, therefore, how essential it is that my application to you be treated in confidence, whatever its outcome.

I have immense experience on every side of the trade/industry. I worked my way up to my present position of ... from the very bottom.

I am 29 years of age, happily married and have three children. My working experience has been varied. The last eight years have been spent as ...

If there are disadvantages in your background, then even these may be turned to account.

Although I have no practical experience of the ... industry, the many years I have spent in ... and ... have provided me with a knowledge of potential markets which I believe you would find valuable.

I started work immediately I left school. Happily, evening classes enabled me to acquire the necessary academic background. My experience has largely been obtained through practical work in the industry. It stretches over 17 years and into every aspect of the business.

Note that the approach is positive. You do not start with an apology: 'I regret that I have no academic qualifications ... a minimum of practical experience ... no knowledge of your particular trade ...' This is dangerous. There are rare products that are sold so softly that the 'spiel' begins in an apologetic tone. These do not include yourself. Show both honesty and an appreciation of your difficulties by including your demerits. Every really good job application combines the maximum of self-praise with the minimum of immodesty.

Avoid the following:

If I am to be accused of immodesty, then so be it. I must tell you that I am highly qualified for the job.

I do not wish to appear immodest, but ...

Modesty forbids me to set out at length the full scope of my experience.

I do not wish to indulge in self-praise, but ...

False modesty is insincere. Apparent insincerity is death to the job hunter. If you cannot even market yourself tactfully, how can you manage others, sell products or organize a business? So away with the pretences. Tell the truth about yourself, with confidence.

No one likes to buy rejects, unless, of course, they are very cheap. Everyone on the other hand, likes to think that a product is custom-made for them. Those who think they have beaten the queue – and the market – to their purchase are contented and satisfied buyers.

Selling your house or your business? Then obviously you would not wish it to appear to be a drag on the market. Lines which 'stick' are sold off cheaply in sales. Conversely, the best way to encourage a customer to buy is to indicate that the goods are in short supply. The retailer who puts a 'sold' sticker on an article is sure to have enquiries: 'Can you order one of these for me? Have you any more of these in stock?' One bright spark disposed of a line of really slow sellers by putting a large sign on the counter which read: 'Sorry – only one per customer.'

Your application must not appear shop-soiled. You must not be an apparent reject.

The higher the position you seek, the more important it is that you appear to be 'just the person we need'. While you are asking for the post, you must quietly indicate that the job's crying out to be done by you. The sales assistant shouts: 'Last few only ... hurry and buy before stocks run out ... I offer you a special, unrepeatable rate ...' The customer feels privileged, contented, eager to buy. That is the feeling that your letter must instil in your potential employer.

Do include a *curriculum vitae* – separate from but attached to your letter. Make sure that key parts of it are in the letter itself (as in examples already given) and that you introduce it with care. Thus:

> To avoid overloading this letter with details, I am enclosing an account of my background and experience.

> I hope that the enclosed CV will be helpful.

> I have prepared for you a brief account of my experience, background and personal details, which I enclose.

In no circumstances should the 'enclosed details prepared for you' appear to be photocopies. The extent of your lie becomes all too apparent. (Please do not laugh – I have seen this sort of error perpetrated dozens of times by job applicants who remain applicants always.) Personalize even the documents that you send out to all and sundry. Reangle them – retype them, remodel the words, the introduction, the ending. Above all, be selective in the facts that you present.

Selectivity is the key to successful self-marketing. Study your market. Read the advertisement. Find out what the company does. Put yourself in the place of the selectors. Ask yourself: If I were them, which details would I want? Which would I regard as unnecessary? Which facts would impress, which would depress? What should I include and what leave out?

You cannot always know the precise nature of the work you are applying for, or even of the business. Before you write, try to do some basic research.

I used to select staff for an organization which included in its name the words 'Bridge'. It had nothing to do with card games, but every batch of applications contained at least one in which the writer praised his or her own skill as a card player. They brought a touch of hilarity to the selection board, but never an interview. Our advertisement was clearly worded, so the applicants' interpretation of it showed lack of either care or intelligence or both, and disqualified them for an executive position. If in doubt, they should have had the good sense to prevaricate or to check by telephoning before committing themselves to paper.

Now, there is a phrase which is so often used that it has lost its true meaning. A pity, because it expresses it admirably. When you put your application on to paper, you commit yourself.

When you do come up for interview, you are almost certain to be asked why you want the job. Work out the answer before you apply for that interview or you may never be given one. Ask what they want of you and see how you can best indicate that you are able to fulfil their needs. You are writing a sales letter, but you have special difficulty because your product is yourself.

How extraordinary it is that the same people who will spend many useful, thoughtful hours drafting letters for products or services will not spare a few minutes to do the same for their career prospects.

35

Interviews for jobs

An interview is a one-sided negotiation which both sides initially hope may lead to a contract of employment. There are no laws which regulate the carrying out of that interview, but its outcome should always be confirmed by letter.

With luck, both sides will be satisfied. The prospective employer will offer a contract and the prospective employee will accept that offer. In any event, courtesy requires the employer to inform the interviewee of the decision taken. Interviewees who reject the job should write and say so.

If the employer has offered expenses, then by coming to the interview the interviewee accepts that offer and those expenses must be paid. Otherwise, the law does not require prospective employers to pay for the interviewee's fares or meals.

Invitation to be interviewed

We shall be pleased to interview you for the post advertised. Would you please call at the company's above address and see our HR Manager, Ms Roberta Jones, at 12.15 pm on Monday 18

July. If this time is impossible for you, please telephone Ms Jones and arrange another mutually convenient appointment.

Application accepted

Thank you for attending the interview. Congratulations. We are pleased to accept your application for the position of ... I shall write to you in detail within the next few days.

Application refused

I regret that it has not been possible to offer you the post for which you kindly applied. We were deluged by applicants, many of whom had more experience in the field than yourself. We thank you for your application and wish you every good fortune in the future.

Applicant rejects offer

Thank you very much for offering me the post of ... After anxious consideration, I have decided not to change positions at the moment.

Applicant rejects, again

My company has made me an offer which I find irresistible. In the circumstances, I have decided to stay on in my present post. Thank you for your kind offer and for the time which you gave me.

36

Retirement and after

Retirement may mean exile to some seaside town or sheltered village. Too many professional and business people, though, are put out to grass when their appetite for work is undiminished and their desire for rest is as small as their need for money is great. They leave one position to hunt for another.

Many of the letters in Chapter 34 can do good service irrespective of the user's age. Here are some more suggestions.

> I am retired, but not retiring. I have twenty years' experience of ... and I wish to continue to put it to good use.

> My company operates a compulsory early retirement scheme, so as to make way for the young. For this reason, and for this reason alone, I am now seeking alternative employment. My interest in my work and my ability to carry it out are equally undiminished.

> I am in far too good health to give up working, and I have several new schemes for improved production/data processing/personnel selection/which I am looking forward to putting into effect.

> Since my compulsory retirement three months ago, I have had the opportunity to undertake refresher courses in computer programming, management methods and general business administration. These, combined with my forty years of active

experience in the industry, make me anxious to begin my new career without delay. I would be happy to call on you.

I am as physically fit as I am mentally restless, and I want to continue my work in our industry.

My younger colleagues and I worked in complete harmony. A rule which the company made inflexible in order to provide incentive and opportunities for them has resulted in my being forced into early retirement. Your company, I understand, is more concerned with mental agility, physical energy and commercial experience than with age.

My years of experience being great, I trust that my years of age will not disqualify me from the post you offer. My expectation of active working life is at least another ten years. By then, I hope to have made a thoroughly firm and useful impact on your ... department/company.

It is true that I shall bring with me years of age but at the same time, I have an accumulation of knowledge and know-how, skills and experience which are probably unique in the industry. These have been acquired, as you know, in the service of ... plc. Owing to their compulsory retirement scheme, I would greatly appreciate the opportunity of talking to you, and of discussing some of the ways in which I would hope to be of long-term service to your organization.

Please would you see me? I have some extremely interesting information which I would like to discuss with you. I am desperately upset at having been forced into early retirement, but the knowledge and know-how, skills and contacts acquired in the service of my former company would, I believe, be of great potential use to you.

Your letter must indicate active, youthful energy plus great experience. The very defects of age which have caused your dismissal may be your greatest assets. Avoid the miserable old phrases – 'I am a youthful 65', or 'I am an active, middle-aged person'. Be positive.

Are you applying for a position at a lower salary than you were previously enjoying? Then do not say: 'My needs are now much smaller. I am prepared to accept less than I previously earned.' The former allegation is probably untrue and the latter is all too obvious.

Anyway, why should you not start climbing the ladder once again? Many of the world's most successful people achieved their eminence longer after companies retire their executives. De Gaulle, Churchill, Adenauer, Eisenhower, Reagan ...

37

To the press

Letters to the press are a democracy's safety valve. It may vent your frustration to write to the press, but it is also pleasant to see your views in print. Here's how:

1 Be concise. In spite of all the rubbish that is published, the extraordinary fact is that nearly every paper (trade or professional, local or national) is short of space. A rambling epistle will be spiked.
2 Avoid defamation (see Chapter 48). You may make 'fair comment on a matter of public interest'. You are entitled to express your opinion, but any facts upon which it is based must be substantially correct. The opinion need not be 'fair' in the sense that it is reasonable or sensible. The most outrageous views are entitled to (and often are given) an airing on the letters page.

Most editors are anxious to keep out of trouble with the laws of libel. One Sunday national rejected (and paid me for) a commissioned article on the ground that, although accurate, it might cause offence to people with whom they were anxious to keep on friendly terms. When I gently suggested that they were meant to be '*the* fearless paper', the then features editor replied, 'We must have a paper to be fearless in!'

3 The swifter your reaction to the news, the more topical your piece, the more cogent your reasoning, and, of course, the more highly respected and well known your name or that of your organization, the better your chances of getting published. The letter itself makes news.

Another suggestion. It is often worth telephoning the editor or letters page editor of the paper and asking whether a letter on the subject you have in mind would have any reasonable chance of publication. There is rarely any guarantee that your letter will appear but if it has a blessing from the top the chances are good.

For the paper's opening and closing preferences ('Dear Sir/Madam' as opposed to 'Dear Editor', or perhaps, 'Dear John' – and 'Yours etc.' as opposed to 'Yours sincerely', etc.), see the column. If you want to sell (even for nothing), study your market. Fall in with the idiosyncrasies of the paper.

You are not, of course, expected to accept the views of the editor. There are two main sorts of letters arising out of editorial policy and a paper normally publishes examples of each. The first (well-loved, naturally) is in sympathetic praise, and designed to encourage a continuation of the pressure and publicity already given. The other presents the other side of the coin.

> My company is deeply involved in the Smoke Town Development Scheme. It is vital for the future of this entire area. With it, adequate and varied employment is almost guaranteed. Without it, the stagnation from which this town has been suffering for a considerable time past will be exacerbated. Young people will continue to leave the area in search of better jobs. This in its turn will deepen the recession and depression which have in the past done much to keep new industry away.
>
> While appreciating the disadvantages which you so succinctly explained in your editorial, we are surprised that you do not appear to welcome the development as a whole. To find the local newspaper fighting local progress is sad indeed. We would at least like your readers to know that all criticisms of the development are most carefully considered; that we make every effort to produce schemes which will cause the least possible disturbance and the maximum advantages to the amenities of the neighbourhood; and that (commercial advantages apart) we believe that when this development is complete, the changes it will bring to the area will be welcome to all.

Knowing the fair hearing which you give to those whose views differ from your own, I do hope that this letter will be published. We are anxious that your readers should understand that there is another side to the case.

Note:

1 Remember that the editor always has the last word, so vituperation may not only keep your letter out of the column but also provoke a reply in kind, which may be more harmful to you than your aggressive approach was to the paper (or to the case which you were seeking to attack).

2 As in all aspects of life (business and private), aggression and hostility provoke like response. Attacks which are reasonably subtle, as well as civil and well reasoned, generally produce better results.

3 The letter on page 124 would be greatly strengthened if a paragraph could be added setting out (in brief, concise sentences) some of the facts on which the writer relies, to show that the development will help the area, and/or that care has been taken in its preparation.

In support of editorial policy

It was a pleasure to read the clear and concise explanation of the aims of the Smoke Town Development in your editorial. In the long run, the capital which will be attracted to this area will benefit not only the industries which will be represented on the new estate, but everyone in the county. A revived local economy combined with new and well-paid employment will bring money into the shops, work to those who provide services of all kinds and satisfaction to the local householders of every category. That, certainly, is our wish and that of all others concerned with this project.

Enclosing letter to editor – supporting paper's policy

We were delighted to read your editorial in your last month's issue and hope that you will find it possible to publish the enclosed letter in support. The development has so many detractors that were it not for your lively and helpful support, we would be very pessimistic about the future. Thank you, in any event, for your guidance and encouragement.

We would be very grateful if we could have 2 000 reprints of the editorial concerned. How much would they cost? If you could kindly get the appropriate person to telephone our Mr Jones, this would be much appreciated by us all.

With renewed thanks and hoping to see you again soon, and with all best wishes.

Note:

1 A covering letter often helps. If you know the editor, so much the better. Even if the letter which you want to have published is an attack on editorial policy, a covering note can do no harm (see the example immediately following these notes).

2 Who should sign the letter? In general, the more weighty the writer's reputation, the more likely it is to be published. If the writer of the covering note is not the person who has signed the letter, explain why. (Again, see next example.)

3 Remember the reprints service that papers usually provide, willingly and at a low price. Reprints are often the most helpful, influential and cheapest form of public relations material.

Enclosing letter to editor – attacking paper's policy – from someone known to the editor

I am taking the liberty of sending you a letter from our Chairman, which we all very much hope that you will publish. We appreciate that it contains an attack on one aspect of your editorial policy, but we know that your shoulders are broad and that you are seldom unhappy when your editorials provoke a lively reaction! We would have preferred, of course, to have written in support of your policy, but know that you will not take it amiss if we hope that your views will not be immutable and that words of our Chairman may influence not only your readers, but (dare we hope) even yourself?

Anyway – and seriously – we would all feel much better if the other side of the case could be ventilated.

Meanwhile, my kindest regards and apologies for troubling you. With best wishes.

Note:

1 This letter is suitable for an executive on friendly terms with the editor. The editor will, of course, realize that the letter signed by the big boss was probably drafted by a public relations officer or executive.
2 Do not forget to end with a friendly greeting, preferably handwritten.
3 Intelligent antagonists make a clear distinction between their regard for the editor and the paper and their poor view of the opinions they wish to attack.

Enclosing letter to editor – attacking paper's policy – from a stranger

I enclose a letter from our Chairman. I do hope that you will manage to publish it. There is extremely strong feeling here that it would be fair to give space to the other side of the picture.

I am asked to tell you, also, that if you or any of your staff would care to meet us, we would gladly make arrangements for a site visit. We feel sure that we could provide information and help which would assist you in what we appreciate is a difficult task.

Note:

1 Find out and use the editor's name. This distinguishes the covering note from the 'Dear Sir' letter for the column, and subtly flatters the recipient.
2 Editors are busy people, so the shorter and more pointed your remarks, the more likely they are to reach the boss's desk, rather than that of an assistant. Then, perhaps, your letter will reach the person in charge of the letters page with a recommendation from the top.

Used properly, the letters page is an excellent way to make your views known, to influence decisions and to get publicity for you and your organization. But in every newspaper there are many other pages. They too provide you with opportunities.

A survey conducted by *The Guardian* in May 1996 came to this conclusion: 'A considered estimate would put the amount of PR instigated material in a broadsheet paper at 50 per cent. The figure is higher in the local press and in tabloid nationals.'

That means that over half the stories you read in the papers are there, not through the efforts of journalists fearlessly searching out the truth, but because individuals and organizations placed them there.

When you look at the scope and scale of British journalism, that figure begins to make sense. There are 24 national newspapers, 95 daily provincial papers and some 430 weekly paid-for papers, 950 freesheets and 12 000 specialist publications. And they all need copy.

You have read about how to have your letter published in the letters page. When you write to a newspaper or magazine with a view to receiving coverage in the news or features pages, similar advice applies. You must make your story attractive, interesting and newsworthy.

The Public Health Laboratory Service in the Midlands wanted local businesses to know more about the work it did and the services it could provide. The Laboratory director wrote to the local press. He didn't include a long list of the epidemiological and microbiological tests the lab could provide. He simply offered to write a 'Bug of the Month' column. The editor loved the idea, the column became a great success and the Laboratory gained a great deal of free publicity.

Your letter has to have a hook. Have you ever seen this headline in a newspaper? EVERYTHING GOING WELL. PEOPLE GENERALLY HAPPY. No, papers need conflict, issues and new ideas. They need what journalists call 'an angle'.

A newly established company wanted coverage for the training courses it held for professional witnesses – the specialists who are called on to give expert evidence in the civil and criminal courts. The marketing director wrote to the broadsheet papers offering to do a piece on the advice his company would give to O.J. Simpson.

The result? Full page coverage in both *The Times* and *The Observer*. At that time, the O.J. murder trial was receiving massive publicity. Journalists were desperately looking for ways to 'refresh' the story, to find a new angle. A feature on witness training provided them with exactly that.

Think hard about your story. Decide on your hook. Choose the most appropriate newspaper or periodical. Make life easy for the journalist by suggesting a new slant or a fresh idea and you can secure valuable editorial coverage for your organization. And with a full page, black and white advertisement in a national newspaper costing around £40 000, that must be worth a letter.

38

Congratulations and condolences

Sincerity is the keynote of the good personal letter. Slush is unpleasant for the feet in snowy times and revolting to the mind whatever the occasion. Here are some precedents of the few, appropriate, welcome words that a situation of joy or sorrow requires.

Congratulations

Well done! Myra and I were delighted at your good news. We wish you every good fortune.

Well done – but formal – on promotion

I have been asked by my Board to tell you how very delighted they were to read of your promotion. They wish you every success, and so do I.

On honour

Together with all your colleagues in the industry, we rejoice at your new distinction. We wish you many years of good health in which to enjoy it.

Recovery from illness

We were all delighted to hear that you are back in harness. Congratulations! We hope that you will now keep fit – and that you will resist the temptation to overwork.

Please join me for lunch, as soon as possible. Meanwhile, best wishes from us all.

Condolences

We were shocked to learn your tragic news. Your husband was a magnificent colleague and a man whose opinion, company and judgement we all valued. We shall miss him.

We all feel very helpless, but if there is anything that any of us can do to be of assistance, we would regard it as a favour if you would not hesitate to let us know. Meanwhile, my colleagues and I send our warmest regards and our most sincere sympathy. We share a deep feeling of loss, and trust that you will be spared any further sorrow for many years to come.

What can I say? We were all so very fond of Mary. I know that even though she had been suffering for so long – and to that extent, her passing must have been a merciful release for her – it must still have been agony for you to lose her. We would all like to be of help to you, if we can. Is there anything we can do? Please phone or write or call. We really would like to do something constructive, if possible.

It occurs to me that you might like some help with the legal miseries of winding up the estate or dealing with personal effects. If we can take any of these worries off your hands, please tell us and we will put the company solicitors at your service. Their probate department is very efficient.

Janet joins me in sending you our fond sympathy. We hope to see you soon.

The sad news of your husband's passing was received at Head Office today, and on behalf of the directors and the staff I send our sympathy to you and to your family in your loss, the sadness of which we share.

From the time your husband joined our company, he earned everyone's respect, and we have lost a valued colleague and friend.

As some measure of tangible help at this time, I am pleased to confirm that John was a participant of our Staff Life Assurance Scheme. In due course, you will receive a cheque for £ ... from the Trustees of this fund.

Our insurance company will need a copy of the death certificate and probate of your husband's will. If he left no will then Letters of Administration will be required. Your solicitors will advise you how to apply for these if that course is necessary, but if there is any help which I or my assistant, Mary Brown, can give you, please do let me know.

Note:
The formula for letters of condolence should be:

1 Sympathy.
2 Comfort – which includes words of praise for the deceased.
3 Offer of practical help – if possible, in concrete terms.
4 A touch of normality – including suggestions for future meetings, and even sometimes a small touch of humour.
5 Tact – which generally includes the avoidance of emotive words such as 'death'. 'Passing' is a fair substitute. Usually, you need not mention the circumstances giving rise to the letter. They will be only too painfully obvious. Do avoid pomposities like 'sad demise', 'tragic passing on' and 'the world to come'.
6 If you know that the survivor to whom you are writing holds strong religious beliefs, then write a letter such as that which follows. Otherwise, the words may be regarded as tactless (at best) or cant (at worst).

The comforts of religion

I know that Bob's life was given up to the service of other people – not least through our church. I am sure that the world in which his spirit now lives for ever will be one free of pain, where his good deeds, fine character and remarkable unselfishness will receive their reward.

Meanwhile, Jenny and I send you our most sincere sympathy. We would like to be of help in some way and as we are coming up to town next week, we will telephone to see if we can drop by for a chat. If we can be of assistance before then, we would be very pleased if you would contact us.

I need hardly say how shocked and upset we are at the news – but we are confident that, with God's help, you will find strength.

Our fondest greetings to you.

Note:

Never criticize the deceased, however well deserved such criticism might be!

39

Thank you

Call it a 'bread and butter' letter, if you like, but there is none more important. People who consider that they are entitled to be thanked but who receive no words of gratitude may feel both angry and hurt, as may those who feel they have not been thanked adequately.

Thank you letters need not be long. They must be sincere and apt. They should preferably be handwritten (see Chapter 28) when they are personal – such as thanking the recipient for hospitality. But for business occasions they may be incorporated at the start, and probably repeated at the finish of an ordinary, routine letter.

Here are some thankful openings and closings:

> My husband and I are very grateful to you for your hospitality, which we greatly enjoyed, and which we now look forward to returning.

> We cannot thank you enough for the way in which you and your wife put yourselves out to make our visit to ... so happy and memorable.

> We did enjoy the hospitality of your home, the company of your family and friends and the theatre evening which you arranged

for us. We hope that it will not be long before you and your wife visit . . . Our home will then be yours. We hope that you will feel as at home in it as you made us feel in yours.

It was a delightful lunch – which I enjoyed. I am grateful to you for your time and hospitality.

Many thanks indeed for your courteous kindness to me when I visited your works.

My Chairman has asked me to say how greatly he appreciated the courtesy extended to him when he visited your factory. He will himself be writing to your Chairman, very shortly.

Your letter was immensely appreciated. Thank you so much.

It is no exaggeration to say that, thanks to you, yesterday was the most memorable day we have had for a long time.

I cannot resist the opportunity of expressing once again our appreciation for your courtesy and kindness.

We are very grateful to you for your help, which went far beyond the call of duty.

You have done us a very good turn – and we look forward to the chance of repaying your kindness.

We do appreciate your help and are grateful for it.

You are a good friend and we all appreciated your help.

It is really a delight to work with people who are not only colleagues, but also excellent friends. Thank you for . . .

It is a pleasure to compete against you! Your courtesy and consideration last night were enormously appreciated.

Finally, we would like to express once again, our appreciation for . . .

In conclusion, we send our renewed thanks for . . .

With our renewed thanks and all best wishes.

Thank you once again for . . .

Next, who better to thank than a grateful customer or client?
Thus:

> The Managing Director has asked me to write to you to say how
> grateful he and the other members of the Board are for your
> kind remarks.
>
> We always do our best to ensure the satisfaction of our cus-
> tomers, but it gives us great pleasure when we receive letters of
> appreciation such as the one which you were kind enough to
> write.
>
> My directors very much hope that our business relationship
> with you will continue for many years, to our mutual satisfac-
> tion. We are passing on your kind sentiments to ... and the staff
> at ... Thank you again for your trouble and thoughtfulness in
> writing.

So far, we have looked at letters of thanks for favours received.
But offers may also be acknowledged with gratitude. Refusal
may need to be both tactful and graceful. Here are some helpful
formulas:

> It was extremely good of you to ask me to ... I am very upset
> that I cannot accept.
>
> I have tried hard to put off a previous engagement for exactly
> the same time as yours ..., but it was impossible. So I must
> refuse your invitation – but I do so with much regret and hope
> that you might ask me again another day.
>
> My trouble is an overfull diary. I so wish that you had asked me
> just a few days ago. As it is, I have a prior appointment which I
> cannot now cancel.
>
> How very good of you to invite us! And how very sorry we are
> to have to decline.
>
> Once again, I fear, duty must come before pleasure. I cannot
> accept your invitation, because on that very day ...
>
> As our American friends put it, could I please take a raincheck
> on your invitation? I simply cannot get away from the
> office/works/factory/shop at the moment.

I do hope that you will not be offended at yet another refusal. Somehow, our meeting seems to be fated not to take place.

It is very kind of you to ask my wife and myself to visit you at home. But we both feel that this time the hospitality should be ours. As it happens, the date you mention is very difficult for us. May we instead suggest that we would welcome a visit from your wife and yourself to us on . . .

I am always happy to add profit to pleasure. Your suggestion of a business lunch next week is one that I would accept with alacrity, were it not that . . . As it is, please forgive me – perhaps our secretaries could fix another day, convenient to us both?

Then what of the guests who do you a favour by coming? The speakers or lecturers invited by you who give their time because they are fond of you, or need your goodwill? Or the colleague who works overtime to help you replan your works? Or maybe your son or daughter, who deserves a pat on the back for some business kindness? All too often we forget that those closest to us are still entitled to our gratitude and that if they earn it, we are very lucky.

Here are some useful lines of appreciation:

You fill your time with service to others – and we are both proud and privileged that you spared your afternoon last week, to visit our . . .

We are extremely grateful to you for speaking to our staff/board/sales representatives/sales conference/management trainees. You will have noticed how attentive they were to your words. Perhaps the greatest tribute of all was the bombardment of questions, which you so skilfully answered and parried. It was with great regret that I had to conclude the meeting. We all hope that you will come again. Meanwhile, we do thank you.

No father is entitled to favours from his daughter/son – and that I receive so many from mine really gives me enormous pleasure. Apart from being an honoured offspring, you are a good friend to your aged dad and I am grateful.

I do really appreciate, Father, the confidence you have placed in me. I am also grateful for the financial security which you have now given me. I shall do my utmost not to let you down.

I would like you to know how grateful I am to you for the help you have given to me in this very difficult period. We are now through the woods. I do not know how I would have survived it if it were not for your support.

Finally, a word of thanks may serve as an important reminder. For instance:

Just a note to thank you very much for sparing me so much time during my recent interview. I greatly look forward to hearing from you.

It was very good of you to agree to ... We all look forward to your visit on ... I confirm the details, which are ...

Thank you so much for saying that you would send me ... This will be immensely helpful.

It was very good of you to promise ... Your support/ help/ action will make all the difference.

40

Appeals

A successful appeal must have direct and personal relevance to the reader. The wealthier the recipients, and the greater their reputation for generosity, the greater the number of such letters they will receive. Benevolent millionaires have told me that they are sent hundreds of appeal letters every week. Naturally, most of the 'round robins' are consigned to the basket. Most replies must say 'No'. The appeals that produce results are usually personal, from people whom the donors do not wish to disappoint or whose requests they respect.

At best, the response results from true altruism. The cause is worthy and the giver is willing.

Often, however, self-interest plays its part. People like, for instance, to gain a touch of immortality for themselves or their families through the naming of a building or a bed, or an inscription in a book or on a roll of honour.

Then there is money. To earn you must spend. Those who send forth their bread upon the waters of charity may harvest some commercial gratitude. Call it sordid if you like, but when the chairman of your main customer asks for a donation for his pet charity, can you refuse?

Anyway, you have your own pet project, haven't you?

Doubtless the day will come when you will write to a friend or colleague who raided your pocket, saying, 'I am sorry to be a nuisance, but the cause is excellent.' They will sigh, reach for their pens, and return the compliment. They are presidents or treasurers of their charities, you of yours. The cynic may have little use for either of you. But the organizers of the charity and (far more important) its beneficiaries, will bless you both.

We all want blessings and need them. Faith, hope and charity are the bastions of our world. The doing of good deeds and the giving of charity lie at the root of every religion, and even those who are irreligious may respond to non-secular appeals, just in case.

Successful appeal-makers have much in common with sales people. They must study the market and frame and angle their letters accordingly. Each has his or her own methods. Study those of the successful and copy them. Before you criticize, consider the results.

A famous *schnorrer* – a Jewish beggar by a much kinder and more appreciative name – is said to have approached a leading Rothschild. 'Will you help me?', he asked.

'You know that I do my best not to refuse help,' replied the charitable magnate. 'But I do feel that when you come to see me, you might at least wear clean and respectable clothing.'

The man looked down at his shabby garb and then up at his prospective benefactor. 'Mr Rothschild,' he said gently, 'Do I presume to tell you how to run your bank? No. Then please do not tell me how to beg!'

It pays to learn from others. Make a collection of the appeal letters, brochures and circulars that you receive. Ask yourself, 'Which ones strike home to me? Which have meaning and vitality? Which make me give?' When you know that, your research is becoming productive.

Usually, you will find that the letters which are really appealing combine sincerity, simplicity and personality. Their presentation is sufficiently unusual to remove them from the ruck. These will be enclosed with a personal note.

As I expect you know, ever since we discovered that David had severe learning difficulties, Mary and I have thrown ourselves into the work of the Happiness Home. Unfortunately, the place

is desperately under-staffed and cannot cope with even a frac-
tion of the children who need help. The greatest shortage is
money. Please will you make a donation/join our
organization/take an advertisement in our brochure *(or as the
case may be)*? If you wish to contribute from a charitable source,
the Registered Charity Number of the Home is ...

We do know the many calls that are made on your generosity,
but this one is very close to us. Thank you so much.

Or:

This is my year as chairman of the Trade Benevolent Fund. The
need is enormous, as the enclosed pamphlet shows, and the fund
does a vast amount of good. I hope that you and I will never
need it, but many others do. I enclose a circular about the ways
in which you could help and I know that you won't let me
down.

Or:

We are desperately short of resources for our work. Please
would you help? I look forward to hearing from you.

Or:

Forgive this personal approach, but I do know the great interest
you take in the Trade Benevolent Fund. If your company would
help us this year, we would be immensely grateful. Your last year's
contribution was £X 000. As we ran up a deficit of £Y 000 during
the twelve months just ended, we are asking our contributors to be
good enough to increase their donations. Would you please help in
this way? The Fund does magnificent work – and it could not
operate without the assistance of the leaders in the trade.
With all best wishes.
PS I have wrung the necks of my own Board, and I am happy
to say that my organization will be doubling the contribution
this year and donating £X 000, to set the ball rolling.

A few more rules, then:

1 The best fund-raisers are those who give – of themselves and
of their money. They win by example.
2 Ask the recipients of your letter to write back to you personally – so that they do not think they will get away with
sending a miserable token to some anonymous appeals orga-
nizer.

3 Never forget to send covenant forms, and/or state the charity's registered number and/or, if appropriate, to ask for Gift Aid.

If ever you are stuck, find someone who is a successful fundraiser, ply him or her with some judicious flattery – possibly washed down with food and wine – and ask for help. Collecting money for charity is a very big business: it is highly competitive and requires the best organization. Nothing can beat a devoted angel at the helm.

Finally, a word from the head of a very generous charitable foundation: 'When I receive beautiful brochures on splendid art paper, I get cross and the charity generally gets nothing from me. If it has that sort of money to throw away, then it cannot be as short of funds as it says.'

The converse, from a rival philanthropist: 'You must spend money to raise money. Unless you have a good-looking, well produced and professional set of literature, I will not believe that you are a well and professionally run charitable organization in which I should invest.'

There must be a happy medium somewhere.

41

Occasions for letterwriting

It was an occasion of dread boredom. The dinner was bad, the company mediocre and the speeches dragged on towards midnight. Only one person at my table looked engrossed and happy. He was making notes of the speeches on the back of his menu card.

'Quotable material?,' I asked him, when the evening eventually reached its morning end.

'That's what you were meant to think,' he said. 'Actually, my wife is abroad and I was writing her a letter!'

I learned my lesson from this. When I am out speaking in public and someone scribbles at my side, I am not flattered. The diner or listener is probably writing a letter.

I have since taken a leaf out of the diner's menu card. Agendas and minutes, blank at the back, provide a monstrous temptation to scrawl a note of affection to far-off friends, or even a skeleton draft of a letter:

> I know that you will forgive the paper, but at least it will indicate that I really am spending the time when you are away immersed in miserable business!
>
> You will see from the enclosed that where I am I wish I were not! Anyway, there is nothing like a touch of complete after-

dinner boredom to provide the incentive for me to drop you a
note, to thank you/remind you/to communicate . . .

Of course, there are limits to the people who would regard this
sort of note as a compliment. The same style and stationery that
is ideal for close relations may destroy good relations with
those who expect a more dignified approach.

Next time you are sitting on the platform listening to dull
speeches or enduring wretched boredom at any sort of gather-
ing pick up your pen and the nearest available scrap of clear
paper – and write.

Part V

Supplies and Systems

42

Stationery and supplies

Your letterheads, notepaper, compliment slips and other stationery are your literary shop window. If they are scruffy, inartistic or ugly, they will produce a bad impression and harm you or your business. If they are neat, well laid out and handsome, your correspondence is far more likely to have its desired effect.

Why, then, do so many able people devote so little time and thought to their stationery? 'Oh, just order in some ordinary letterheads. Same as last year will do. Don't bother with proofs.' Fatal. If you want your letters to be telling, then they must be typed or written neatly and accurately on well designed and competently printed paper.

If you are in business, do you spend heavily on advertising agents, buying space in the media? Then why not spend a comparatively small amount on a professional designer to plan your stationery, or to redesign your logo? You would not begrudge the money if you were preparing a sales leaflet. So pay to sell yourself.

Cut the time spent by your staff and yourself on correspondence. The use of good precedents should help. Save through improving your word processing and other systems of work or

149

administration, but spend just that small amount more on the paper on which your letters are written.

These suggestions apply, of course, to every sphere of both business and personal correspondence. They are especially relevant to letters concerned with the law. If you want credit to fend off legal proceedings brought by your debtors, then appear prosperous – which is impossible if you write on cheap paper. You wish to attract top management? Then give the impression that you operate a top outfit. Make sure that the design at the top of your paper will advertise your modern and business-like approach. You would not allow your staff to look slovenly, when dealing with your clients or customers, would you? Then smarten up the stationery on which they present your written case.

43

The envelope

Appearances count and first appearances count most. The scruffy, disreputable, uninteresting, poorly presented envelopes that first greet the recipients of too many letters do the writers no credit.

'I can't spend a fortune on envelopes of pristine white,' you say, trying to justify the miserable brown things in which you send out your mailshots. Nonsense. The top mail-order companies make sure that even if every letter is mailed by the thousand, each recipient feels that he or she is the one for whom it is intended (see Chapter 33 for more on direct mail).

Naturally, there are limits to this doctrine. Nobody feels better about receiving a bill just because it is in a splendid cover. Accounts should be sent out as cheaply and expeditiously as possible. But deal with letters differently, even if they are requesting payment of accounts. They are part of the front of your business.

For mass mailings, sticky labels can be useful. Use these computerized products for extensive list mailings, especially where the secretary or clerk will open the envelope and the recipient is unlikely to see it.

Where your letter is personal, have the envelope neatly

151

typed. Note the form of title for the recipient (Chapter 8). Check that the name is properly spelt, and the address accurate.

If your letter is intended to be personal, mark the envelope. If you do not want the executive's secretary or the director's assistant to read the contents, mark it 'Personal and confidential' or 'Strictly personal and private'. Underline that in red, if you wish. (For the legal implications of opening employees' mail, see Chapter 53.)

If there is no indication that the contents are for the eyes of the named person only, you should not be surprised if anyone else in the business happens to open and read the letter. The great do not sort and sift their own mail. If your letter is intended for the personal attention of the addressee, then the envelope must say so. What is written on the envelope and the humble cover itself both matter far more than most letterwriters realize.

44

Files and filing

There are several reasons for writing a skilful letter. First, you want it to have the appropriate effect on the recipient. That's the top copy. But the duplicate copy may be equally important. It will jog your memory about the contents, and it may be used in evidence, either at a hearing or in advance to show the strength of your case and so prevent a court action. Copies matter.

Of course, the finest, cleanest copy of the most brilliant letters is useless if you cannot find it when required. So you need the most efficient filing system you can set up and operate.

Apart from the general rule that the simpler the system, the more effective it is likely to be, my advice is: devote time, thought and care to creating or adapting a system to suit your own requirements. Commercial enterprises which depend so much on the accuracy of their records may spend millions on the development of their products but begrudge the few thousand pounds needed to establish a first-class arrangement for their files. Of all business economies, this is the least intelligent, if only because a really good system will keep the filing itself to the minimum.

Simplicity, as always, is the key. At best, each letter will carry the reference number of the file and/or the reference of the matter in question (Chapter 5). If it carries neither, then the secretary or typist should be asked to put at the top right-hand corner of the copy some identifying number of reference which will make the finding of it child's play. If a bright youngster cannot operate your filing system, consider changing it.

Electronic storage

Storing documents electronically rather than on pieces of paper inside files is both a blessing and a curse. Whether a document is saved on an individual PC or on a shared storage area what seemed to the author a logical name to call it and place to store it is not always obvious to anyone else trying to locate it subsequently.

Therefore, to take full advantage of the benefits of electronic storage you must devise a systematized and approved file naming and saving structure that all your organization or individual work groups will adhere to. Failure to do this will result in chaos and could cost your business a lot of money.

To help you in this task there are a number of computer packages designed to help you manage and organize your documents electronically. There is also web-based technology that will automatically index your entire computer storage area to enable you to search for documents by subject or content.

A fuller description of the storage options is given in Appendix 2.

45

Notes

Making and using brief notes can save the letterwriter a vast amount of time. If you write or word-process your correspondence, you can cast your eye back over the lines and amend, reamend, excise or add as you see fit. When you are dictating on to a tape, it is annoying to go back further than the last sentence or two. Notes are the best answer.

The best system? Jot down at random ideas you wish to convey. Then put them into logical order. A one-word note should spark off a paragraph of dictation. The notes create the skeleton of the letter. Keep your notes concise. There may be some key sentence that you need to work out in detail, but if there are many of these you would be better off dictating them (in logical order if you can) and then, if necessary, redictating from the first complete draft.

The first objective of notes is to help you to structure a clear, concise and accurate letter incorporating the ideas that you want to convey. Their second purpose is to jog your memory. They need not be written at the same time as the letter. Carry a notebook or diary and practise jotting down ideas as they come to you – otherwise they will disappear, often never to return.

These notes can be incorporated into those you make shortly before you write your letter.

In practice you will find the need for notes gradually lessens as your experience and skill increases. Once you are used to putting your notes into logical order and dictating them in letter form, you will find that you can sit back and put your thoughts into proper order without writing them. The more flowing the letter, and the more polished your writing, the more likely it is that you will have mastered the art of using simple notes as indicators for complex thoughts.

You often need to make notes at meetings or even at meals to remind you of arrangements made with some other party. Sometimes this can be left to an attendant secretary or scribe. Often you must do it for yourself, particularly if the meeting is private or informal. Always make a note at the time: accuracy is important, and so is concentration.

How do you achieve precision in note-taking? Ideally, by using shorthand. The art of swift writing is too often regarded as the preserve of secretaries or professional shorthand writers. It is an enormous asset to everyone.

Even the simplest abbreviations help. Know the gramma-logues – that is, tiny signs used for the most common words – and you can cut down your writing time by a quarter. Then leave out the vowels and another 25 per cent disappears. The words 'and', 'of', 'but', 'if', 'company' and the like, occupy an inordinate amount of space in ordinary script, so do 'public', 'private', 'board', 'director' ...

A few hours spared to learn even the most basic shorthand can save hard and unnecessary labour, and add to the accuracy of your performance.

Learning to type or to word-process is more difficult and time-consuming. If you can handle a typewriter or PC, you can bash out your own correspondence. Most executives used to leave this side of the business to the secretary or typist and many still do. But note-taking (preferably in shorthand) is a necessary chore for all. A good note prevents many a bad letter.

These suggestions apply to every field of professional and commercial letterwriting. Barristers or solicitors who are able to take careful notes of a judgement, in shorthand which they can read and transcribe, are blessed. Architects or surveyors who are able to jot down the requirements of their clients swiftly and accurately will not only save time, but will also please their clients, whose precious moments are also spared. Doctors use their own form of shorthand for most of their reports. So should you.

You may even be able to train your secretary to read your shorthand. Why not? Plenty of secretaries read back each others' notes. Teach your shorthand to your children at college, and instead of writing to you once a month, they will be able to write once a week with precisely the same enthusiasm (or lack of it). They will also be able to make notes in classes and lectures with maximum speed. For this they will bless you – and blessings from our children are not to be spurned.

46

Dictating and dictating machines

When the typewriter and shorthand came in, the art of calligraphy went out. As executives' responsibilities increased, so the labour of committing words to paper was passed on to secretaries and shorthand typists.

Then along came the dictating machine and the word processor, and audio typists took over a great part of secretaries' work, while shorthand typists disappeared. So, consider how to make the best use of all these aids to letterwriting.

Naturally, you need spares for your equipment. Strangely, the same business people who ensure as a matter of course that spare parts are available for their works machinery are often incredibly mean when it comes to their own office equipment. Most systems are well engineered and reliable, but from time to time they collapse. To operate a successful word processor or dictating machine system, you must have 'swops' to hand. And of course you need swift and reliable service engineers.

Shop around very carefully before you plump for any particular make. Once you have chosen, you are probably landed with it forever. Nothing is more aggravating than to have different machines in the same office or group of offices. Secretaries or copy typists fall ill. Then tapes have to be sent to others for transcription. Machines break down. Then you will want only one set of spares.

Which equipment should you choose? All depends upon your needs, your pocket and your preference. Recommendation is important. Experience should be a good guide. For ideas, try purchasing the various office equipment journals, sent in most cases at no cost to you.

So much for the basic equipment of the letterwriter. Now for some hints on its use.

There is only one way to make a dictating machine your ally and that is to use it. At first, talking into a microphone and flicking a switch for a replay seems strange, but it is well worth the effort.

It is essential, of course, to have the right equipment, in decent working order and adequately supplied with spares.

Whatever machine you use, you will soon acquire your own technique. Writing a letter by machine takes practice. You will find that your ideas gradually fall into proper shape as you go along.

If the letter is complicated you should certainly jot down notes (see Chapter 45). If you are blessed with shorthand, you may want to rough out the letter fairly fully and then dictate it. In many cases a precedent book may help your dictation to be smooth and unworried. At worst, remember that it is far better to dictate a draft and hack it about after typing and redictate, than to use longhand. Apart from the reduced physical effort involved, the result is generally more satisfactory and almost invariably less time-consuming. And what have you to offer that is more valuable than your time?

Techniques of dictation vary. Some people dictate even the punctuation. Others leave it to the transcribers. Some dictate slowly, others at great speed. Some are blessed with transcribers who can type swiftly and consistently from tape. Other typists are slower.

You should speak clearly, and give precise instructions. The nature and extent of these will depend on the skill and experience of the person who listens to your words. For instance, you should be able to say simply: 'Please see the precedent at page . . . of the precedents book and adapt it to be sent to Messrs Smith & Co . . . The amount involved is £300 . . .

Omit paragraph four ... Sign off with a paragraph inviting him to lunch ... and send my greetings to his wife ...' Or: 'Please use precedent No. 85 in the Letters Book – but you sign it on my behalf ...'

In some cases you will have to spell out words. 'I spell Rumpleforth ... R-U-M (as in mother) –P (as in Peter) -L-E-F-O-R-T-H.'

To be sure that you have clearly expressed your meaning say, 'I repeat ...', and then respell. If you do not include the words 'I repeat', you cannot blame the transcriber if the same words are typed twice.

One of the greatest risks in using a tape is that the contents are so easily erased. It is irritating to have to redo correspondence because, having forgotten to remove the tape from the machine, you have rewound and then redictated over it. It is even worse if the tape goes astray and never gets typed. The entire office frantically plays back every reel or cassette they can lay their hands on in a desperate (and probably vain) attempt to trace the missing one.

If you operate dictating machines, you must find some system to indicate when a tape has been used, and put it into its box with a slip or sticker indicating the contents, and perhaps the date.

Part VI

The Law on Letters

Introduction to Part VI

If you write letters, especially in business, you cannot avoid potential or actual contact with the law. The chapters in this Part set out the laws that you are most likely to encounter and explain how best to deal with them.

First come those laws which apply specifically to letters: the legal implications of signing a document; the possible effects of carelessness in what you write; copyright; defamation; and other legal traps. Next, the branches of the law which the letter-writer most commonly needs – for instance, contracts in general and contracts of employment in particular. Finally, there are the vexed questions of suing for money you are owed; problems in court; and letters in dispute.

If you follow the legal rules set out in these pages, they should help you to keep as far away as possible from courts and tribunals.

47

Copyright

The ownership of a letter passes from the writer to the recipient. The letter you receive becomes your property. You may file it or tear it up, treasure it or give it away, but you are not entitled to copy it. Copyright remains with the writer. You cannot include the letter in your memoirs without the writer's permission. The laws of copyright are complex. Here is a summary.

Copy one person's work, goes the old saying, and that's cheating. Copy more than one person's efforts, and that's research! As far as the law is concerned, though, if you copy something in which copyright subsists, you are liable to be sued for infringement of copyright – and the more people's work you copy, the more potential plaintiffs you are creating.

In daily office practice, we all work from precedents. Indeed with modern, sophisticated equipment, copying has become very easy. But it has its perils. So consider the Copyright Designs and Patents Act 1988.

Where copyright subsists in any work, you are in general only entitled to copy that work with the licence of the owner. If you invent your own precedents, write your own advertising material, draw your own plans, diagrams and maps, or create your own instructions to your staff, you have the right to prevent them from being copied by others without your consent. Conversely you are entitled to reproduce the original brain-children of others with their consent.

Copyright subsists in 'literary, dramatic, musical' and 'artistic and graphic works'. These terms cover just about everything, from railway timetables through drawings, maps, charts and plans to poetry and literature of the highest order. During the author's lifetime and for seventy years from the end of the calendar year in which the author dies, reproduction is usually only allowed with the author's consent. Nor need the reproduction be exact in order to amount to a 'copy' – a 'copy' is that which comes so near to the original as to give to every person seeing it the idea created by the original. So not only are exact reproductions covered but so are 'colourable imitations'.

Now, copyright means the exclusive right to copy the work, to issue copies to the public, to perform or show (in the case of music, plays and films) the work, to adapt it, or to license or authorize anybody else to do any of these acts – usually in the United Kingdom or countries with reciprocal rights. But who is 'the owner'?

In general, 'the author of a work is entitled to any copyright subsisting in that work'. But there are exceptions – notably under Section 11 of the Act. 'Where a literary, dramatic, musical or artistic work is made by an employee in the course of his employment, his employer is the first owner of any copyright in the work subject to any agreement to the contrary.'

Before the 1988 Act, there was another important exception – where a photograph, portrait or the like was commissioned for payment, the person who commissioned it became the copyright owner. Now the new law gives copyright to the photographer or artists, unless agreed to the contrary. But the subject of the photograph or portrait has an important protection: where the work was undertaken for private and domestic purposes (like wedding photographs), the person commissioning the work has the right to prohibit publication.

All this is subject to agreement to the contrary. If you decide to have your office rebuilt or redesigned and you commission architects to draw up the plans, they normally retain the copyright. The fact that you commissioned and paid for the work is irrelevant. You are not entitled to take their designs, reproduce them and sell them – or even to reproduce and keep them. They *have the right to decide* who may and who may not copy the work.

But there is nothing to prevent you from employing architects on the basis that you will acquire the copyright in their drawings – nothing, that is, except possibly their reluctance to agree.

Equally, if you employ talented people in your office on the basis that they will have copyright in any original works they produce in the course of their employment, the Act does not remove that right. Like so many pieces of legislation, it only applies in the absence of some agreement to the contrary.

What of employees whose literary creation belongs to their employers? Authors now have a new and significant right – a 'moral right' as it is described by the Act – to be identified as author whenever the work is published commercially, and the same is true of musical compositions broadcast in public. But this important right does not apply to one category of authors or composers – those whose works are owned by their employers because of their contracts. So employees are no more protected than before.

Chapter II of the Act lays down how copyright is infringed. If, without the licence or authority of the copyright holder, you copy the work (with limited exceptions for private research), or issue copies to the public, or perform, broadcast or adapt it.

There are exceptions. Libraries have special privileges. So do those who use extracts from works for the purpose of reviews or criticism or for the setting (and answering) of examination questions. In general, though, the rule is simple – literary, musical, artistic and architectural piracy are not permitted.

48

Libel and slander

It is defamatory to publish anything about other people which would tend 'to lower them in the eyes of right-thinking people'. You must not bring others into 'hatred, ridicule or contempt'.

To defame someone in writing or in some other permanent form (including incidentally, a statement made on radio or television) is a libel. To speak ill of another is slander.

The fact that a statement is true does not prevent it from being defamatory, but no one is entitled to a good name which he or she has not earned. So, if sued for a defamatory statement which you can prove to be true, you may plead 'justification'. You may claim that the statement was substantially true. The effect of a plea of justification is to repeat – even more loudly and publicly – the very same defamatory statement that you made before. Therefore, if a plea of justification falls, your offence has been severely aggravated. The damages awarded against you will be greatly increased.

A much more helpful defence is 'qualified privilege'. The law recognizes that certain statements may be made for the public good. People must be entitled to speak their minds. Hence 'privilege'.

No action in defamation can succeed in respect of any

statement made by anyone in a court of law. 'Absolute privilege' also applies to all statements made in Parliament. However malicious, untrue or unjustified a statement made in court or in Parliament, it can never give rise to a successful defamation action.

Similar privilege attaches to occasions upon which the law recognizes that the writer of the statement has a public or private duty to make it, and the reader a direct interest in receiving it. For instance, references are business necessities. For that reason the givers of references are protected. They are under a moral duty to speak their minds to the inquirers (although, note, they have no legal duty to supply the reference). The recipient of the reference obviously has an interest in knowing its contents. The occasion is 'privileged'.

Or suppose that you have to write to a colleague about a possible sacking. Your letter alleges that the person was dishonest ... slovenly ... disobedient ... stupid ... unfit to be in your company. He or she is defamed. But clearly, this sort of letter *must* be written. It is essential company business ... The occasion is 'privileged'.

But while privilege in courts and Parliament is 'absolute', when you write or speak to colleagues or supply references the privilege is 'qualified'. The qualification? If the statement was made out of 'malice', the privilege evaporates. 'Malice' simply means some wrongful motive. If it can be shown that the object of making the statement was to harm the person defamed rather than to assist the management in reaching a sensible conclusion, or the prospective employer in deciding whether or not to employ an applicant, the privilege goes. The law is not designed to shield the spiteful.

There is another defence for the writer of evil words: 'Fair comment on a matter of public interest'. So your words must be a statement of *opinion* and not of alleged *fact*. If they were partly opinion and partly fact, then in so far as they are fact, they must be substantially correct. Comment to your heart's content, but do not misstate facts.

The comment must be 'fair'. This does not mean that the recipient or reader or the person referred to must consider it reasonable. In practice, this word exercises little restraint on your comment. Provided you are not simply using the occasion to forward a private grudge rather than to comment on a matter of public interest, you should have nothing to worry about. But do not confuse fact with fiction and, under the guise of comment, propagate false statements about your enemies.

Writers, then, should watch their words, whenever they are writing evil. Remember the three little monkeys? The one with his hands clapped firmly over his mouth is the most intelligent of all. Speak no evil and you need fear no action in slander.

As usual with the law this also applies in reverse. If you are at the receiving end of unkind words, apply these principles and you will know whether, in theory at least, you might have a good action in defamation against your defamer. Do not be surprised, though, if you are advised by your lawyer not to sue.

Defamation proceedings are perilous and unpredictable. Even if you win, you will endure worry, annoyance and expense.

The rich, the determined and the fortunate sometimes succeed, and are very handsomely recompensed for their risk by the court's award. The less well off should be wary before they take that risk.

My advice: with rare exceptions, do not sue. Let the evil words flow over you and they will soon be forgotten. Only lawyers are guaranteed to win lawsuits.

49

Sedition and other traps

There are various other ways in which the law interferes with the freedom of speech or of writing. They are all rare in practice, but still require a weather eye from the letterwriter. So here is a miscellany of civil and criminal consequences which can arise out of use of the wrong word.

First, perjury. If any person who is 'lawfully sworn as a witness or as an interpreter in a judicial proceeding wilfully makes a statement material in that proceeding which he (or she) knows to be false or does not believe to be true...' he or she is a perjurer and may be imprisoned for up to seven years or fined an unlimited amount – or both. So when appearing before any 'tribunal, court of person having by law power to hear, examine and receive evidence on oath', mind what you say. And remember that written sworn evidence must be as accurate as you can make it (see Chapter 50 on affidavits).

Although there are some prosecutions for perjury, when you consider the number of perjurers it is obvious that the fear of committing this offence has about as little effect on the dishonest witness as the terror of purgatory.

The offence of 'sedition' embraces all those practices, whether by word, writing or deed, which fall short of high treason but directly tend or have for their object to excite dissatisfaction or discontent ... to create public disturbance, or to lead to civil war ... to bring into hatred or contempt the sovereign or the government, the constitution or the laws of the realm ... to excite ill-will between different classes of the sovereign's subjects ... to incite people forcibly to obstruct the execution of the law ...' and so on, and so on.

In theory, this offence might put a heavy rein on free political discussion. But in practice it, too, is almost as dead as the proverbial dodo.

Only slightly less dormant is 'criminal libel' – which covers any defamatory publication which the publisher can show to be true *and* published in the public interest. This ancient offence has been said to be justified by the need to stop outraged gentlemen taking up their swords to avenge the attack on their reputation. Like the duel, it has largely slipped into history.

Now for some civil results of uncivil words.

As we saw in the last chapter, defamation may lead to trouble. But has it occurred to you that to speak ill of people's *goods* may be defamatory of their persons? Suppose for instance, that you say: 'Jones is turning out really shoddy stuff these days and selling it at a very high price.' You are hardly heaping compliments on Jones – you are saying, in effect, 'That man Jones is a rogue – he is selling low-quality goods at a high price.'

Apart from libel and slander, words may themselves give a 'cause of action' if they cause damage to a person 'in the conduct of his affairs' or are calculated to cause him pecuniary loss.

Suppose, first, that any sort of property is up for sale. Someone 'without lawful motive' untruly writes that the property is charged, or that there are liabilities upon it, or that the vendor is not in a position to sell. This is 'slander of title'.

Again, if you write that someone is selling goods in infringement of copyright or patent, you may be committing 'slander of title'. But nowadays there are various statutory remedies available to people accused of this sort of behaviour. For instance, section 70 of the Patents Act 1977 says that a person who is threatened with proceedings for infringement of a patent may bring an action for a declaration that the threats are unjustifiable, claim an injunction plus, if any loss has been suffered, damages.

Again, falsely and maliciously to disparage the quality of someone's goods may create a 'cause of action' – if the disparagement prevents their sale. By all means indulge in 'mere trade puffery', but 'knocking' may lead to trouble.

Accordingly, where a false statement is made maliciously (out of a desire to injure and without lawful authority) and produces as its direct consequence 'damage which is capable of legal estimation', an action may lie for slander of title, slander of goods 'or other malicious falsehood'.

Finally, a note on 'malice'. 'Maliciously' has been defined as meaning 'without just cause or excuse'. Unlawfully and intentionally to do 'without just excuse or occasion' an action which causes damage may lead to trouble. But it is certainly malicious to act out of some improper or dishonest motive or with the intention of causing injury. Where there is 'a distinct intention to injure the plaintiff apart from honest defence of the defendant's own property', an action may lie without there being any defamation as such. (For 'malice' as affecting the defence of 'qualified privilege', see Chapter 48.)

Therefore, if you improperly or dishonestly attack the title or property or products of your competitors, they may have a good claim against you. The law approves of competition but frowns upon the more unpleasant forms of 'knocking' the goods and property of others.

50

Affidavits and oaths

An affidavit is a statement of fact, sworn by the 'deponent'. If on conscientious grounds, witnesses decline to take an oath, then they may make the appropriate declaration. What matters is that the court then has documents made by witnesses presumably as a result of careful thought and, if necessary, research. They undertake to tell the truth. If they lie, they are perjurers.

Far too many affidavits are made carelessly. Suppose, for instance, that you want 'summary judgement'. You have issued a writ; there is no apparent defence (or 'triable issue') so your solicitors ask the court to give swift, sharp judgement, without the need for a trial. An affidavit will have to be filed, verifying the facts in the writ and saying that there is no defence – or no defence to that part of the claim in respect of which summary judgement is sought.

Affidavits may be sworn by solicitors. It is then for them to make sure that they are satisfied as to the truth of their assertion or as to the 'information and belief' to which they depose. But the solicitor may provide you with the draft document and ask you to take it to a Commissioner for Oaths and have it sworn.

Even if the solicitor has had the document handsomely and apparently permanently typed on special 'engrossing' paper, this should not deter you from perusing every word with care. After all, morality apart, the other side may not cave in and the case may reach trial. If it does, the odds are that you will have to give evidence. If you end up in the witness box, you will be cross-examined on the basis of your sworn affidavit. A mistake can be very expensive indeed. Take care what you say. Mind your words – particularly when they are written and sworn.

51

The effect of a signature

In business, a handshake may no longer be an enforceable bond. But a signature still means much.

If you sign a letter put before you by your secretary or assistant or executive and (because you rely on his or her honesty or skill or judgement) you sign without reading it, please do not think that you will afterwards be able to avoid the legal results of that letter, at least in so far as third parties are concerned. You will be bound by it. That is the most important effect of the famous decision in the case of *Gallie* v. *Lee*.

Mrs Gallie was a woman in her eighties. She had a trusted nephew named Walter Parkin, who had been kind to her over the years. He was sole beneficiary under her will and a few years before she had decided to give him her house: 'Everything I possessed belongs to him,' she said. He wanted to raise money on the house and she was willing for him to do so, provided that she could stay in it during her lifetime.

Unfortunately, Mr Parkin had a friend called Lee, who needed money to pay off *his* creditors. On the advice of a crooked solicitor's managing clerk, Lee arranged for documents to be drawn up by which Mrs Gallie would sell the house to him for £3 000. Lee would not pay this, but he would mortgage the property.

Lee then prevailed upon Mrs Gallie to sign the document. Unfortunately, her glasses were either mislaid or broken and she did not read what she was signing. 'What is it for?,' she asked Mr Lee. 'It is a deed of gift for Wally for the house,' Lee replied. She signed. Lee paid her nothing. The solicitors obtained a £2 000 loan for him from the Anglia Building Society. Lee raised money on a second mortgage, but defaulted on the instalments. The building society sued for possession. Mrs Gallie and her nephew pleaded *non est factum* – that the document was not hers.

Mere mistakes in the contents of a document you sign will never allow you to avoid its effect.

What, then, of Mrs Gallie's gift?

'A man who has failed to read a document and signs it should not be allowed to repudiate it as against an innocent purchaser,' said Lord Denning. 'His remedy is against the person who deceived him.' Even if he could plead fraud or mistake against the immediate party (that is, the person who induced the signature of the document), he would not be able to avoid the consequences 'when it had come into the hands of one who had in all innocence advanced money on the faith of it being his document, or had otherwise relied on it ...'

The principle? 'Whenever a man of full age and understanding who can read and write signs a legal document put before him for signature which on its face is intended to have legal consequences, then, if he does not take the trouble to read it but signs it as it is, relying on the word of another as to its character or effect or contents, he cannot be heard to say it is not his document.'

Lord Justice Russell agreed. Mrs Gallie, he said, had intended to sign a document divesting herself of her interest in the house. This she had done. Here was no case of *non est factum*.

Therefore mind what you sign. And if people try to say that they did not realize what they were signing, then just refer them or their solicitors to the case of *Gallie* v. *Lee** and that should be the last you hear of that bad argument.

---

*The Court of Appeal's decision was affirmed by the House of Lords, where the case was known as *Saunders* v. *Anglia Building Society*.

52

Negligence

If someone else suffers damage as a result of a careless statement, it may lead to trouble all round. That was one effect of an important decision of the House of Lords.

A well-known merchant bank was asked for a reference. The inquirers wished to know whether a certain company was worthy of credit. The bank supplied the information; and when this turned out to be incorrect the inquirers lost their money. They sued the bank, claiming that although they (the inquirers) were not customers and the information was supplied gratuitously, the bank still 'owed them a duty of care' – that is, was under a duty to them to exercise such care as was reasonable in all the circumstances to ensure that the information given was correct.

'Nonsense,' retorted the bank. 'We supplied the service at no charge and you cannot expect us to have the same liability to you as we would have had if you had been a customer or we had charged you. And anyway,' they added, 'there was a disclaimer on the reference saying that it was given "without responsibility" on the part of the bank or its officers.' And they denied negligence.

The trial judge held that they *had* been negligent. They were under a duty of care, even though the service was given gratuitously. This decision was eventually upheld by the House of Lords, but they also ruled that the effect of the disclaimer was to let them off the hook.

The basic principle was established long ago. We each owe a duty of care to our 'neighbour'. A 'neighbour', in this sense, is any person who we ought reasonably to anticipate would be likely to be affected by our negligent act. If, then, you are a manufacturer, you have a liability in contract to the people who buy your goods. If the goods are faulty, then you are in breach of contract. If you are negligent and they suffer injury, loss or damage, then you may be held liable.

But your responsibility does not end there. It extends to 'the ultimate consumer'. Suppose that you manufacture drink. It must be obvious to you that the person who is likely to drink it is not the wholesaler or retailer to whom you actually sell the stuff. The 'ultimate consumer' – the customer of the retailer or caterer – is the person who will be poisoned if the drink is defective. He or she is the 'neighbour' of the manufacturer.

Thus there is a liability in the law of negligence not only to those whom you know but even to complete strangers.

'The bank,' said the House of Lords, in effect, 'must be taken to have realized that the reference was asked for with a purpose in mind. The intention was that the reference be acted upon. Therefore the bank ought to have realized that if the reference was incorrect, the result might well be that the recipient would suffer damage. The bank owed a duty of care to that recipient, even though the service was given gratuitously.' The milk of human kindness may prove a very costly commodity.

Negligence had been found against the bank and a duty of care was owed. The damage was also proved. That left the disclaimer. The bank had given the reference upon the explicit and clear understanding that it was not to be held responsible for the accuracy of the document. The recipient could not go behind that disclaimer, which was fully effective in protecting the bank. As a result, the House of Lords did not have to consider the question of whether the defendants had been guilty of negligence. The bank escaped because of its disclaimer.

Since that decision,* many business people have shivered

*Hedley Byrne v. *Heller and Partners* (1964).

slightly and taken insurance cover. The givers of every sort of reference must take care not only to avoid defamation in circumstances in which malice may be imputed to them (see Chapter 48), but must also be careful to ensure that, if asked for a reference for one Peter Smith, they do not provide it in respect of another. They owe a 'duty of care' to the recipient – and to Peter Smith.

The case of John Lawton[†] demonstrates this point. After he was made redundant, Mr Lawton used his former employer's name when he applied for a new job. They took up his references with a form of questionnaire, which was duly answered; and his new employers immediately dismissed him because the answers were so unfavourable. Somehow Mr Lawton obtained a copy of the reference. He felt it was so inaccurate that it showed 'culpable negligence' by the HR manager; so he sued.

'Even if we were negligent,' replied the firm, 'and even if we were liable to the new employers, we cannot be made liable to the man himself.' But the court disagreed: 'There is ... no doubt that the plaintiff relied upon the defendants to give an accurate opinion and state accurate facts in the reference.'

Unfortunately for John Lawton, the court also found that there was no negligence, because there was enough evidence for his ex-employers to have reached their unhappy conclusions about him.

Not only employers writing references need to take care. What you write – or even what you say – when trying to persuade people to work for you may be quoted against you in court. Mr McNally[‡] applied for a job on an oil refinery in Libya. He was interviewed by Mr James, on behalf of the firm. Mr James gave him a number of assurances about the job. Mr McNally signed a contract. The small print said that if he did not pass various tests, his employment could be terminated. In the event, he lasted ten days in Libya before being dismissed with (eventually) wages in lieu of notice. The job was not one for which he was in any way qualified.

Not content to sue the firm for leading him thousands of miles under false pretences, Mr McNally sued Mr James for negligently misrepresenting what was on offer. He won

†*Lawton* v. *B.O.C. Transhield Ltd.* ‡*McNally* v. *Welltrade International Ltd.*

substantial damages. Despite the defendant's argument that there was no more than a 'moral duty to protect the ambitious employee from himself', the judge ruled: 'The questions here are whether Mr James took it upon himself to advise the plaintiff whether he was qualified for the job, and whether as a result of that advice the plaintiff was induced to enter into the contract.' The answer to both questions was 'yes'.

It is not enough to prove that you were negligent in giving the advice or information concerned. To obtain damages against you, your correspondents – like Mr McNally – would have to prove two other things: first, that the statement concerned was acted upon; second, that they suffered damages, foreseeably arising from the negligence.

Infallibility being a divine attribute, everyone in business makes mistakes. Happily, most of them lead nowhere too disastrous. Indeed the great advantage of making a mistake is that next time you may recognize it. If others do the recognizing, then that is unfortunate. But it is only if they do not realize that you have been in error and actually take action as a result of your mistake that they will have a legal remedy.

Suppose, for instance, that you make a misleading statement in a letter. As a result, the recipient consults his or her board, solicitor, accountant and management consultant and then – bolstered by expert approval – takes action along the lines you have suggested. The chances are that they could not blame you. There were too many intervening people, facts and ideas.

Alternatively, suppose you make some provocative statement. It may never enter your mind that anyone would be stupid enough to act upon it without further research or inquiry. But maybe you were being obtuse. The question is: would the 'reasonable person' have expected you to have foreseen that your correspondent would act upon your words? Should you reasonably have prophesied, had you applied your mind to the situation, that your words would give rise to someone else's action? If not, then your mistake will lead nowhere – at least so far as you are concerned.

Assume, now, that the recipient of your letter can overcome both these hurdles. He or she has still not reached the end of the trail and must show that the damage was not 'too remote'. Take an example from an ordinary road traffic accident. Your employee caused it through careless driving? You are responsible, as if you yourself had been at the wheel, if the driver had the accident while working in your employment. Therefore, you would have to compensate anyone who suffered injury, loss or damage as a result – provided that this was foreseeable.

The cost of repairing the other vehicle, or reimbursing the injured person for lost wages, or damages in respect of personal injuries – all these can be laid at your door.

But suppose, on the other hand, that the other driver missed an important appointment and hence a potentially profitable contract. That would be his or her misfortune. That damage would be 'too remote'.

All this involves some very complicated legal considerations. If your letterwriting leads to the threat of legal action, the sooner you contact your solicitor, the better. Meanwhile, treat this chapter as a warning – and take care.

53

Responsibility for letters

Of course you do business on paper. You cannot write all your own letters, all of the time, can you? Even if you insist on signing all your mail, this chore probably comes at the end of the day, when your mind is weary and your soul longing for hearth or home, bed or bottle. You make mistakes.

Or are you a chauffeur-driven executive, signing mail in the back of the car? Then what if the driver lets his or her mind wander and negligently causes a collision?

Whether the employee's mistake is made in words or on wheels, the employer may not be the only sufferer. But who will be liable to pay? Consider the legal rules on 'vicarious liability' – when the sins of the employee may be laid at the door of the employer.

Every employer is liable to third parties in respect of damage caused by employees within the scope of their employment. If you have the benefit of someone's work, then you must accept the burden of his or her mistakes. On the other hand, the mere fact that you employ people does not mean that they are bound to you, every long hour of the day. They are free to sail off on independent frolics of their own. When they do so improperly, they are outside the scope of their employment and their employers bear no responsibility for their misdeeds.

Your secretary may operate a typing bureau from home. If she does so on her own account and in her own time, then she is responsible for her own negligence. You normally bear no liability.

Or maybe your driver crashes a car while off on an evening's gambol. The fact that this happens in the company car will not place liability on the company.

Was the negligent employee acting 'within the course of his or her employment', or 'independently frolicking'?

An attendant at a petrol station was forbidden to smoke. He lit a cigarette and threw down the match. The devastating explosion destroyed property belonging to third parties.

'He was not employed to smoke,' argued his employers. 'Indeed he was forbidden to do so.'

'He was employed to put petrol into tanks,' said the Court. 'At the time of the accident, that is precisely what he was doing – although in a thoroughly negligent, improper and forbidden way.' His employers were liable.

Therefore the test is *not*: 'Was the employee doing something forbidden?', but rather: 'Was he or she about the employer's business?' Tell your employees what not to do and (if their misbehaviour is sufficiently serious) you may dismiss them if they disobey. But for third parties, what matters is whether or not the employees were doing their job.

Suppose, then, that your secretary is guilty of a clerical error. Perhaps she leaves a nought off the price of goods offered. At the foot of the document we read: 'Dictated by Mr Smith but signed in his absence.' Poor unfortunate Mr Smith – he should have been there. The company will be bound by the offer just as it would have been if Mr Smith had signed it himself – or, for that matter, if it had carried the company seal.

You are asked for a reference. You hand it over to your HR manager for attention. Negligently, she provides the wrong information – and in breach of your instructions she omits to include a disclaimer of liability? The recipient of the missive suffers damage through relying upon the carelessly written words? Then it will be no answer to the claim to say: 'My employee was guilty, not I'. Her carelessness, committed in the course of her employment by you, is your own. In some ways, husband and wife are still regarded as one by the law – but the employer and employee are united far more often when it comes to the rules on vicarious liability.

A letter arrives at the office, shop or factory. It is addressed to a member of staff, care of the firm. You open it. It contains confidential information. 'Anything that comes to the firm is liable to be opened,' you say. But are you right?

If the letter is addressed to James Smith, c/o Jones Ltd, then Mr Smith must expect to have it opened. The normal procedure in most businesses is for the mail to be opened centrally and then passed out to the appropriate departments or individuals for attention.

If the envelope is marked 'Personal' or 'Confidential', then the employer has no right to open it. Instead of coming to the employee in the course of his or her employment, it has simply arrived at the place of employment. To open it then would be to interfere with Her Majesty's mail. And if by mistake you open a letter not intended for you and discover its highly confidential contents, the law says you may not make use of, or pass on, those confidences.

54

Contracts

A contract is a bargain made between two or more people. It has a number of essential elements. First, there must be an unconditional offer. Second, this offer must be unconditionally accepted. Third, there must be 'consideration' (except in Scotland). Fourth, the required formalities of the law must be complied with – and writing may be one of these. Let us start with writing.

Generally, no formalities are required for a contract to be binding. Most contracts are as complete and binding in law if they are made orally as if every term were written in letters of gold. There are exceptions: contracts of guarantee or for the transfer of an interest in land must be evidenced by some sufficient note or memorandum in writing, signed by the party to be charged. Contracts of hire purchase, of marine or life insurance and for the transfer of shares require writing; so does a contract of apprenticeship. But a contract to buy goods may be made orally, making a telephone conversation being sufficient to wrap up the deal. So, too, a contract of employment (but see next chapter).

Where a deal has not been made or confirmed in writing, it may be difficult to prove its terms. One party may say that one term was agreed, the other something different.

For this reason, confirm your agreements in writing whenever you can. And do make a note or confirm anything important said to you on the telephone. Remember, if a dispute follows and reaches court, you will be allowed to refer in the witness box to notes made at the time.

What about the other essentials? First, the offer. 'I offer to sell you X quantity of Y brand goods.' That is an offer. Simple? Not necessarily. It is, for instance, important to distinguish an offer from a mere 'invitation to treat'. If you advertise goods, you will not necessarily be bound to sell them, either at the advertised price or at all. Basically, the position is the same as when goods are on display in a shop window. They are not 'offered for sale', in the technical, legal sense. Potential buyers are invited to make an offer to buy. Those offers may be either accepted or refused; or a counter-offer may be made – that is, an offer on different terms.

Second, the offer must be *unconditional*. If you say, 'I'll sell you these goods, provided I've enough in stock,' you can avoid completing the deal if you have not enough in stock. That is not an offer capable of immediate acceptance.

Third, *unconditional* acceptance: the offer must be accepted in its entirety. For instance, suppose that you offer to buy goods which you see advertised. You set out the price you are prepared to pay and the dates when you wish to have delivery and you leave no 'ifs' or 'buts'. The supplier writes back saying: 'Thank you for your letter. Your order is hereby accepted and we confirm that the delivery will be made in accordance therewith.' The deal is done.

Now suppose the letter of acceptance contains, printed at the bottom or on the back, terms and conditions inconsistent with your own. In effect, the suppliers are saying: 'We accept your offer – but subject to your agreeing to our terms and conditions as printed hereon.' This is not an *unconditional* acceptance. It is a 'counter-offer' which may be accepted or rejected by potential buyers as they see fit.

Once you realize this, you tend to take a closer look at the terms and conditions on letters or other documents of this kind. Remember that if you do nothing about them and simply accept the goods, the chances are that the counter-offer will be the offer and your acceptance of the goods will be the acceptance. Hence, that acceptance will be subject to the supplier's terms and conditions. An exception to this is exclusion clauses, which may not be valid if they are unreasonable.

But what if the buyer never reads one of the terms in tiny type? Too bad. Provided that it is legible and not excluded by law, it is still one of the terms of the contract. If people do not choose to read contractual documents that is their lookout.

What if they couldn't have understood the terms, even if they had read them? Then they should have asked a lawyer to explain them. Only customers under the age of eighteen or of unsound mind may be able to avoid their contractual obligations.

Generally, the terms of the contract, whether oral or written, are part of the bargain. If that bargain is binding, those terms will be included in it, provided only that these were sufficiently brought to the attention of the contracting parties. (Like the words on the back of a ticket, referring you to the company's regulations.)

Like an offer, an acceptance (if it is to conclude the bargain) must be unconditional. If, for instance, the buyer says, 'I accept your offer of these goods, subject to approval by my directors,' then there is no deal until the directors have given their approval and this fact has been communicated to the supplier.

Now comes the fourth essential: 'Consideration'. This, in English law, simply means some *quid pro quo* – some return for the value or promise given. 'In consideration of our paying you £..., you agree to supply me with...' The consideration 'moving' in one direction is the promise to pay the specified sum; in the other direction, it is the promise to supply the goods in return for that sum.

'The customer should have put down a deposit, I suppose...' Correct. He or she would then have been saying, 'In consideration of my giving you this deposit, you agree to hold the goods at our disposal for the specified period.' If you had accepted the money, you would have been bound to give the customer the option on the goods concerned for the period agreed.

Conversely (the deposit being an 'earnest of good faith' on the customer's part), if he or she had failed to exercise the option and to take up the goods, you would have been entitled to keep the deposit. As usual, you *could* have made some agreement to the contrary. You could, for instance, have agreed to give credit in the same sum as the deposit, if your customer decided to opt not to purchase ...

If a buyer allows one of his servants to place an order on his behalf, then he is bound by that order as if he had given it himself. This must be so or the business world would halt. A company has no existence in human form. Someone must act for it. Even individuals cannot do everything for themselves. If you give someone your authority to contract on your behalf, then you will not be able to avoid the contracts made by that person pursuant to that authority.

'But suppose that the person had no authority ... can the principal then refuse to accept the arrangement made?' That depends. He can do so if the agent had neither his actual nor his 'apparent' authority. Otherwise he is almost certainly bound.

If you 'hold someone out' as having your authority to act on your behalf (if, in legal terms, you give them your 'ostensible' authority) then you are, in effect, saying to other people: 'This person is my agent, entitled to contract on my behalf.' If someone relies upon this statement and as a result makes a bargain with you, then you will be bound by that bargain. You must not 'hold out' people as having authority which they do not in fact possess. If you do, you cannot expect the law to free you from deals made as a result.

Finally, the contract must not be 'too vague to be enforceable'. The law will not make contracts for business people who do not bother to do so for themselves. So sometimes it is possible to avoid a contract if it can be shown that any of the essential elements of that contract are missing.

For example, suppose that your customers had agreed to buy goods. The delivery date was fixed and the goods themselves were decided upon. But you left the price unfixed. Alternatively, suppose that the price was fixed but that the quantity to be taken was not. In either case, one of the essential terms of the deal was missing. The contract would be too vague to be enforced.

To sum up: if a person of full capacity makes an unconditional offer which is unconditionally accepted by some other

person of full capacity and the terms of the contract are adequately set out and agreed between them – and provided that
there is 'consideration' – the contract is complete. Where agents
make contracts on behalf of their principals, it rarely matters
whether the agents had the actual authority of the principals to
contract on their behalf. It is enough if the agents had their
principal's apparent authority to do so. And writing is only
necessary in exceptional cases. When required, it should always
be undertaken with precision and care.

55

Contracts of employment

A contract of employment is an agreement between employer and employee under which the employee agrees to serve and the employer to employ, on the terms stated. A contract of employment (with the sole exception of a contract of apprenticeship) does not have to be in writing to be fully binding in law. But thanks to the Employment Act 1996, the employee must be given written particulars of its most important terms within 13 weeks of the start of the employment or within four weeks of any variation.

The written statement, which may be contained in one or more letters or other documents, must identify the parties (especially important if the employer is a company which is part of a group). It must specify the date when the employment began and state whether or not the employment is continuous with any previous employment (if so, when did that employment begin?). It must state the employee's job title. And it must give the following particulars of the terms of employment as at a specified date not more than one week before the statement is given:

1 The scale or rate of remuneration, or the method of calculating remuneration.

2 The intervals at which remuneration is paid (that is, whether weekly or monthly or by some other period).
3 Any terms or conditions relating to hours of work (including any terms and conditions relating to normal working hours).
4 Any terms and conditions relating to:
   a Holidays and holiday pay (including the manner in which holiday entitlement is arrived at, especially when the employment comes to an end).
   b Incapacity for work due to sickness or injury, including any provisions for sick pay.
5 The length of notice which the employee is obliged to give and entitled to receive to determine the contract of employment.
6 To whom the employee may turn if he or she has a grievance or a query regarding disciplinary procedures – plus details of both grievance and disciplinary procedures or where these may easily be found.

The particulars need not be in any set form and may (and generally should) include other terms not required by law, such as:

- clauses giving employers the right to search employees or their property
- restraint clauses
- clauses warning employees that breaches of the employer's health and safety rules may lead to dismissal.

Employees are (generally, and in broad terms) only entitled by law to written particulars when they have been employed for 13 weeks. Any changes must be notified within four weeks. As for unfair dismissal protection and redundancy pay, to qualify for written particulars employees must work at least 16 hours a week, or eight hours after five years' continuous service.

If the particulars given are disputed, an industrial tribunal has the power to decide who is right, and to 'declare' the correct particulars. Remember, the contract is made when you offer employment and the candidate accepts. Issuing particulars is not an opportunity to rewrite the contract.

The prudent employer makes all employment offers in

writing, with a reference in the letter to the detailed particulars. These may be in the form of a handbook, or a collective agreement with a trade union.

It is worth your time and effort to ensure that employment contracts are correct, because disputes can be very costly and time-consuming. Bear in mind that you cannot unilaterally, without every employee's consent, change what is in their contracts – short of dismissal, offering new jobs all round, and running the risk of a claim for damages for unfair dismissal.

56

Letters in dispute

The more serious or costly the dispute, the more important the correspondence is likely to prove, and the more weighty its probative value. So the courts have evolved a system to ensure that most correspondence, material to an action, is revealed to the other side (if they do not already possess it), and produced for the court's inspection. This process is called 'discovery of documents'.

In any legal action, the parties must set out their contentions in so-called 'pleadings'. In the High Court (which deals, in general, with claims over £50 000), the plaintiffs' 'cause of action' is pleaded in a 'Statement of Claim'.

Next, the defendants put their answers into a 'Defence'. This may include a 'Counterclaim'. The plaintiffs will then file their 'Reply' and 'Defence to Counterclaim'.

In the County Court, an action starts with a 'Particulars of Claim'; and this is succeeded by a Defence (with or without Counterclaim) and a Reply.

If any of the 'pleadings' is obscure or does not set out the case in sufficient detail, 'Further and Better Particulars' may be sought by the other party. If the litigants have not declared their case adequately, there will now be a chance to do so, if necessary as a result of a court order.

The object of the exercise, then, is to enable trial judges to have the contentions of both sides spread before them, so that (on the basis of the evidence, and the law) they may decide the matter one way or the other.

If any document is referred to in a pleading, the other parties are entitled to a copy of it. If, for instance, your claim rests on a letter or an invoice, or on written particulars of a contract of employment, the other parties may demand a copy and on payment of the appropriate copying charges (if any) are entitled to have one.

Moreover, where a pleading does not specifically rely on a document, but there may be some relevant letter, plan, map, order form or what-have-you, upon which the party intends to rely, the other litigant may demand that the document be 'identified'.

Once the 'pleadings are closed', and the contentions of the parties are clearly set out on paper, the time has come for each to reveal documents relevant to the proceedings, which are or have been in their possession. 'Discovery' will take place of all material documents, and these must then be made available for 'inspection'. Litigants are not allowed to keep some useful document up their sleeve, only to produce it with a flourish at the hearing, preferably while cross-examining a star witness on the other side. They must reveal it beforehand.

To make sure this is done, a formal list of documents has to be prepared and presented. Sometimes, the list will be supported by affidavits in which the litigant swears that these are all the relevant documents which are or have been in his or her possession.

Sometimes, instead of a list, the documents will be set out as part of an affidavit. Not all relevant documents must be shown. Some are 'privileged'. For instance, any correspondence which you may have with your own solicitor will be privileged. Nor must you produce an Advice given to you by your own lawyer. But letters which passed between the parties and which were not written 'without prejudice' (see next chapter) will have to be revealed, even if they go against your case.

That explains why the art of letterwriting is such an important adjunct to successful litigation. Careless letters cost cases. Be careful. Mind how you write. Your letter may one day be read out in court. Better still, it may win your case before it reaches trial, which, of course, is by far the best sort of legal victory.

Parliament has added to the legal protection given to victims of poison-pen letters – and of unjustified or malicious threats such as are sometimes used by the unscrupulous to intimidate their debtors. Those who write such letters may now face prosecution.

Your natural reaction may be, 'Good, about time too, but nothing for me to worry about.' But the law is widely, and a little vaguely, drawn, and you *do* need to bear in mind the limits on what you can put in a letter without risking the recipient being able to take the matter to the police. This is what the law – the Malicious Communications Act 1988 – prohibits:

- sending letters or articles which convey an indecent or grossly offensive message
- sending threatening letters
- including deliberately false information in a letter
- sending anything else which is of an indecent or grossly offensive nature.

In each of these situations, an offence is only committed if the sender intended to cause distress or anxiety to the recipient. And it is a defence to show that a threat, if that is the source of the complaint, was used to reinforce a claim which the sender believed he or she had reasonable grounds to make, and the sender believed the threat was a reasonable way of pursuing the claim. For example, a threat of court action if your debtor does not pay up may be acceptable. But it will not be so if you do not honestly believe in the validity of your claim. Nor will it be reasonable even for genuine claims to peddle false information to distress your debtors into paying.

Consumers' representatives are waking up to the possible use of the new law to control some undesirable debt-collecting methods: make sure you are not open to legal attack.

57

'Without prejudice'

What is the effect of putting 'without prejudice' at the top of your letters? If you enter into negotiations which (as they inevitably will) involve concessions on your part, can these be thrown back in your face if the negotiations fail and the case comes to court? How far is it safe to make admissions in correspondence if you put the magic words 'without prejudice' at the top of the letter?

The effect of 'without prejudice' letters was laid down by a judge, many years ago. He said this: 'I think they mean "without prejudice" to the position of the writer of the letter if the terms he proposes *are not* accepted. If the terms proposed in the letter are accepted, a complete contract is established, and the letter, although written "without prejudice", operates to alter the old state of things and to establish a new one.'

Once 'without prejudice' negotiations have reached fruition, the court must be entitled to look at the letters to see whether that allegation is well founded. A judge put it: 'It is no objection that the agreement is contained in letters which are headed "without prejudice".' If you make an agreement, then you cannot say: 'It was only made in the course of negotiations.' If the negotiations had failed, then the court would be entitled to know that there were in fact attempts at settlement, but the letters themselves would remain privileged. Once an agreement is reached, the court must be entitled to look at the letters in which it is contained. If the parties cannot agree as to whether or not the letters contained a binding agreement, the court must examine them if it is to decide.

Mr Justice Ormrod dissented. He said, 'The court will protect, and ought to protect so far as it can, in the public interest, "without prejudice" negotiations because they are very helpful to the disposal of claims without the necessity for litigating in court. Therefore, nothing should be done to make more difficult or more hazardous negotiations under the umbrella of "without prejudice".

> I am well aware that letters are headed 'without prejudice' in the most absurd circumstances, but the letters in the present case are not so headed unnecessarily or meaninglessly. They are plainly 'without prejudice' letters. Therefore the Court should be very slow to lift the umbrella of 'without prejudice' unless the case is absolutely plain.

The judge looked at the correspondence, but decided that the case was not 'absolutely plain'.

The principle is clear. Mark your letters 'without prejudice', so as to ensure that if no agreement is reached your position will not be prejudiced through any admissions or confessions that you may have made. Once an agreement results from the negotiations, they cease to be 'without prejudice'. They are 'open' for all the world to see.

Another case when letters (but not their contents) may be considered is where there is unreasonable delay. In many cases, plaintiffs have been driven out of court because they did not pursue their claims with sufficient vigour. But it would be wrong for claims to be dismissed through delay where the parties were trying to avoid court battle through 'without prejudice' parleys. Allow or encourage your solicitors to haggle. But if they delay when negotiations have ceased, then harry them.

Still, the actual contents of 'without prejudice' letters are only admissible where an offer contained in them has been accepted. For that reason a court will not be allowed to see them merely where they contain admissions or acknowledgments of a debt, where it is alleged that the debt has become 'statute barred'. If the six-year period has passed since the debt was incurred or acknowledged in writing, the fact that there were intervening negotiations will not revive it.

If you want to keep your rights to your leased business premises, under the Landlord and Tenant Act 1954 (as amended) you must serve your notices and if necessary make application to the court within the period specified. If you are kept haggling until that date has passed, your rights will be lost. Whether the negotiations were oral or in writing, provided that they were 'without prejudice', your landlord's position will remain unaffected – unless, of course, you can prove you actually reached an agreement.

The privilege which the law gives to 'without prejudice' letters may be 'waived' by consent of the parties to the negotiations. If all agree that the court ought to see the letters, then so be it. If the writers of the letters wish to waive the privilege, some think that they can do so.

But most lawyers believe that unless both parties consent, or the letters reveal an agreement binding them in law, neither side can improve (or worsen) its position by throwing off the cloak of 'without prejudice' to reveal the contents of their correspondence.

Put your negotiations under the umbrella of 'without prejudice' letters, and if agreement escapes you the chances are that the letters will remain hidden from the judicial eye, now and for ever more.

58

Proof of posting

'We never received your letter,' you write.

'Not so,' comes the reply. 'We can prove that we posted it; you will be presumed by law to have received it; so the offer you originally made was validly accepted by us – and you had no right to sell the goods elsewhere.'

How could your correspondents prove that they did post the letter? If a contract is made by post, is it firm and binding when a letter of acceptance is posted or when it is received? At what stage can you (or your correspondent) still cry off?

Where a contract is entered into wholly or partly by correspondence, and where an offer is made by letter, the deal is done as soon as a 'properly addressed letter containing the acceptance is posted'. Provided the 'offeree' accepts unconditionally and within the time specified in the correspondence, the 'offeror' cannot clamber out of his or her obligations by alleging that he never received the letter.

At this stage, proof of posting becomes important. Ideally, the sender will have taken the trouble to get a receipt for the letter. Recorded delivery will show despatch and delivery and so will registered post.

But most business correspondence is posted in the ordinary

way, without any sort of post office record. Your office should have a posting book, in which the addressee of each letter is noted. Production of the book is not absolute proof, but there will be a presumption of posting which the addressee will find hard to rebut.

'The posting of a letter,' says the law, 'may be proved by the person who posted it, or by showing facts from which posting may be presumed.' Hence 'evidence of posting may be given by proving that a letter was delivered to a clerk who in the ordinary course of business would have posted it.' Alternatively, it could be shown that the letter was put into a box which is normally cleared by the postal staff.

Again, if a letter is properly dated, that date will be taken as evidence of the date when it was written or dictated. In fact, evidence may show that dictation occurred a day or more before. The postmark on the envelope is good evidence of the time and place of posting.

For this reason work out a sensible system for recording of letters posted. And because these rules apply both ways, the stamping of a date of receipt on all incoming post is a very wise precaution.

Note:
The rules are rather different where acceptance of an offer is by telex. Two cases, one of them in the House of Lords – *Brinkibon* v. *Stahag* (1982) – have established that the general rule applicable to instantaneous communication also applies to telex.

Where an offer is accepted by telex, the time of acceptance is when the telex is sent and the place of acceptance will normally be where the telex is received and read and not (as in the case of a letter) the place from which it is sent.

What about fax and e-mail? Probably fax would be simultaneous, like telex. But as with a letter sent by mail, you might have to prove receipt, perhaps through the fax machine's activity log.

As for e-mail, will it be enough to send a message to an e-mail address? Will that constitute delivery? Or do you need to show that the contracting party has read the e-mail? This has not been decided, but the chances are that the rules will be the same as for a letter. You would have to show that the person either received the e-mail or deliberately refused to look at it.

Appendix 1
A guide to style

If you have ever been arrested you will know the police caution includes the words: 'Anything you say may be given in evidence.' It is a solemn warning that we are judged by the words we use.

Similarly, every letter you write tells your readers something about you. A letter can provide your reader with telling evidence about your attention to detail and your passion for excellence. The style and tone of your prose and the way you use, or abuse, grammar and punctuation are evidence of your professional standards and of the values that guide you and your organization.

If, on the evidence before them, you want the jury of readers to find in your favour, it is worth taking time and trouble to write your letters in an accurate, clear and readable style.

This short guide will help you to achieve just such a style. It brings together many of the points made in the main text, and it adds some more tips on making life easy and pleasurable for your reader.

Tone

HM Customs and Excise open their letters with the profoundly uninviting 'To the registered person named above'. This bureaucratic, impersonal tone accurately reflects the balance of power between the organization sending the letter and the individual on the receiving end. The VAT inspector has powers of entry and search and can impose harsh financial penalties. The 'registered person', on the other hand, has no power at all.

When you write a letter, think hard about how the words on the page may sound inside your reader's mind. In that private space, what to you is a simple expression of fact may sound like a cold, authoritarian command. 'The results of the sales promotion are 7 per cent below target. I want your report on my desk by Monday.' Those two sentences are certainly brief and punchy. But what effect might they have on the morale and motivation of the sales executive who has worked night and day to achieve a 93 per cent success rate?

In a letter giving advice or information to a customer it is very easy to write a whole string of imperatives. This is an extract from a letter written by a building society:

> YOU MUST REVIEW THE METHOD OF REPAY-
> MENT OF THE CAPITAL BALANCE, ENSURE
> YOUR REPAYMENT ARRANGEMENTS ARE SUF-
> FICIENT AND MAKE FUNDS AVAILABLE TO
> MEET THE OUTSTANDING PAYMENT.

The customer, who in 12 years had never missed a payment on his large mortgage, was quite rightly offended by the hectoring, instructional tone and by the strident procession of upper case characters. No one likes being 'told off' by a powerful authority with whom there can be no debate or negotiation.

Setting an appropriate tone in a letter is most important. If you are making sure someone knows the health and safety regulations concerning hazardous substances in the workplace it is fine to instruct and inform. But, if you are writing to customers or colleagues, make sure the tone is friendly and involving. And that it makes the reader feel that he or she has some power and importance in the communication taking place between you.

Culture

> We are delighted to respond to your invitation to tender for the provision of services to the elderly, the disabled and the wheelchair-bound in the area covered by the Trust.

This is an extract from a covering letter for a proposal document sent to a National Health Service Trust. The letter almost guarantees the company that sent the letter will not win the contract. Why? Because within the Health Service the acceptable terms are 'elderly people, people with disabilities and wheelchair users'.

This is not an academic debate about political correctness. It is directly connected to business success. The culture of an organization is expressed in the language it uses. To obtain a favourable hearing you must, quite literally, speak their language.

Adjectives – overuse of

An excess of adjectives defeat their own purpose. Simplicity is the keynote to good style and any word that is not doing a job is superfluous. 'As a wealthy millionaire, you will understand the need for financial planning', is an example of saying the same thing twice. The adjective wealthy simply repeats what the noun millionaire already tells us.

Another obstacle to achieving a clear style is adjectival overkill. 'I was shocked to receive your angry, abusive and irritating letter. Your response throughout has been churlish, vindictive, inadequate and irresponsible.' What was the most important feature of the letter – the abuse it contained, the anger it exhibited or its ability to irritate? Which of the four adjectives used in the next sentence describes the worst aspect of the response? In the words of a famous quiz show, 'It's make your mind up time.' A flurry of adjectives serves only to cloud the meaning and reduce the impact of your message.

Mixed metaphors

> Launching at the high fliers first means the grass roots will miss the boat. What's sauce for the goose must be taken on board as sauce for the gander. Miss our target when the starting gun goes and sure as eggs is eggs, our chickens will come home to roost.

At its best, a well-chosen metaphor casts a sudden shaft of light on to a dry or difficult subject. Mix your metaphors and you turn light into darkness. The rule here is a simple one: don't do it!

Collective nouns

These take a singular verb. 'The Government is' not 'the Government are'. There may be 20 people sitting on a committee but you still write 'the committee is' not 'the committee are'. As always in the English language, there are exceptions. You cannot, for example, write 'the police is'. Although you could say, 'the police force is'.

Management speak

Sir Ernest Gowers laid down three golden rules in a guide to civil servants: 'Be simple, be clear, be human.' Badly used, management speak is capable of breaking all three rules in a spectacular way. Imagine receiving this from the human resources department:

> The medium-term strategic objective is to empower customer-facing support operatives to reprocedularize their process goals in a value-driven, mission-oriented macro-environment.

Research proves that people forget the abstract and the conceptual. They remember the concrete and the specific. This example is an impenetrable clutter of concepts and abstractions. Tell someone to improve the way they look after customers by serving them more quickly and there is a good chance he or she will get the message. Tell them to 'reprocedularize their process goals', and they will be massively confused. Which brings us to messages in code.

Messages in code

The DCI wants CID to get the 609 from CRO. The CPS want the TDAs to be TIC'd.

Few trades, professions or businesses are free from the curse of the TLA. Or, to give it its proper title, the three-letter acronym. The community of people who can, without conscious thought, translate the staccato rattle of the TLA into meaning are like the residents of a village. Within the boundaries of the village, each trio of letters is part of a private code, a convenient shorthand for those insiders who are 'in the know'.

For the reader who lives outside the village, the code is both irritating and alienating. If you are not a police officer, you may find this translation of the cryptic message above helpful: 'The Detective Chief Inspector wants the Criminal Investigation Department to get form number 609 from the Criminal Records Office. The Crown Prosecution Service wants the Taking and Driving Away offences to be taken into consideration.'

Long words

Try to avoid long pompous words where there are good shorter equivalents. 'An exhaustive perusal of the totality of the evidential data, particularly that of a circumstantial nature, led the committee to the inescapable conclusion that a thorough promulgation of the new parameters governing benchmark financial best practice and financial malfeasance was obligatory.'

Well yes, Sir Humphrey. Why not: 'The committee took a detailed look at all the evidence and decided everyone must be made aware of the new rules about good and bad financial practice.'

It took Abraham Lincoln just over two minutes to deliver the Gettysburg Address. Two-thirds of the 268 words Lincoln used are words of one syllable. If a great president can inspire an entire nation using short, well-chosen words, just think what the same approach could achieve in your letters.

Brevity

The Lord's prayer contains 56 words. The Declaration of Independence 300. A recent EU order setting the price of cabbage, 26 911.

Capitals

Here is a short extract from one of my favourite documents. It is part of a letter written to every guest of a hotel in Lytham St Annes.

> We Are A 20 Minute Bus Ride From Blackpool So You Are Near Enough For All The Fun And Bright Lights Of The Town But Far Enough Away To Not Have To Endure The Nightlife And Bustle Of A Very Popular Seaside Resort (Unless You Want To Of Course).

The writer is obviously in love with the shift key of the word processor. Making such profligate use of capitals merely devalues the currency of emphasis and importance. The fashion now is to be more sparing. For example, 'Lower Biggleswade's Health Committee' at the first mention but subsequently, 'the committee' or 'health committee'. The word 'Government' is always capitalized if specifically the British Government; not, if not.

The apostrophe

The naturalist and environmental campaigner, David Bellamy, once said that 'It only takes one pylon to ruin one hundred acres of natural beauty.' One misuse of the apostrophe can have a similar impact.

> Its taken two years work. All the product's and all the package's are now in place. All the manager's teams are proud of what theyve done. In the 80's it was all about 'greed is good'. In the 1990's we have proved that thing's are different.

Not so much one pylon as half the national grid. In this example, the writer has forgotten the basic rule: an apostrophe tells your reader that one of two things is happening, omission or possession.

'It's' is short for 'it is' or 'it has' (it's been a cold day). The apostrophe tells you that some letters have been omitted. The correct abbreviated form of 'they have' is 'they've'.

If something belongs to someone, you write 'someone's' or, for example, 'the manager's'.

If it belongs to several people (the managers) and the plural form is made by adding an 's', you write 'managers'' (plural form of manager followed by an apostrophe). Plurals, like people and children, that aren't made with 's', take apostrophe 's'.

Apostrophes do not make things plural. And that includes the 1990s and Boeing 747s. Do not follow the example of the misguided greengrocer by offering tomato's and cabbage's for sale.

Here is a correct version of the above example:

> It's taken two years' work. All the products and all the packages are now in place. All the managers' teams are proud of what they've done. In the '80s, it was all about 'greed is good'. In the 1990s we have proved that things are different.

Slang

In the 1980s, the Government introduced a new form of local government finance. They called it The Community Charge.

Everyone else called it the Poll Tax. For years, subscribers to the telephone system were issued with a Telephone Directory. Which, of course, they referred to as the phone book.

In spoken English, slang almost always replaces the official, formal version. The British Broadcasting Corporation becomes the Beeb. The London Underground becomes the Tube.

In letters, for several very good reasons, slang is rarely acceptable. Using it can make you sound illiterate and poorly educated. Using out of date slang – and there is nothing less cool, groovy and hip than yesterday's trendy terminology – can make you sound old-fashioned and out of touch. Worst of all, slang can offend. Actors, for example, actively resent use of the slang word 'luvvies'.

The boundaries of taste and acceptability are, however, constantly changing. 'Dodgy' has all but made the transition from Arthur Daly to the boardroom. A word originally used only in the phrase 'dodgy motor' is now being used to describe a dubious business proposition or a doubtful credit rating. 'Yuppie' and 'gazumping' have already made the move from spoken to written English. If a word is borderline, the best advice is to leave it out.

Paragraphs

Fowler's *Modern English Usage* defines a paragraph as a unit, not of length, but of thought. With that definition in your mind, take a look at this extract from a supplier of domestic and electrical goods. The public relations department is writing in response to an unhappy customer:

> We are sorry to have troubled you by sending your husband our latest brochure after his death. We have now deleted his name from our mailing list. We are sorry to hear what happened and for any inconvenience caused. However if there is anything you would like to order from our exciting new brochure please do not hesitate to call us.

This paragraph is like a three-piece suite made up of two armchairs and a gas stove. To offer sympathy for the death of someone's husband and in the same breath attempt a sales pitch for a new range of products is inept and offensive.

New thoughts need new paragraphs. Observe this rule and you will express yourself more clearly. You will also make your letters easier to read. By adding horizontal to vertical space you greatly improve the look of the page.

Punctuation

Your best friend is the full stop. Three short sentences are better than one long one with a whole string of subsidiary clauses.

Use commas to aid understanding. Too many in one sentence can be confusing. Use two commas, or none at all, when inserting a clause in the middle of a sentence.

Use a colon to precede a list. 'We must look at three areas: finance, sales and operations.' Use a colon before a sentence quoted in full. But not before a quotation that begins in mid-sentence. 'The Chairman said: "The project is cancelled." But the Finance Director replied that cancellation was "a great pity" and maintained that the decision should be reversed.'

Semi-colons are used to mark a pause longer than a comma and shorter than a full stop. They can help you to distinguish phrases listed after a colon if commas will not do the job clearly. 'The committee agreed on three points: the training project must start within one month; it must be given top priority, in London and in the regions; and a launch event should be held, either in London or Birmingham.'

As always, your first concern should be making life easier for your reader. The modern trend is to minimize the use of punctuation. Very few organizations, for example, now put commas after each line of the address on a letter.

Spelling – American and British

In the early days of its independence, there was much debate in the United States of America as to what this vast new country should adopt as its national language. Some people seriously suggested that, as America represented a new birth of democracy, its citizens should speak classical Greek. It was only by the narrowest of margins that German was rejected. In the event, of course, America chose English. And then proceeded to turn it into something else – American English.

In many ways, American spelling displays a greater logic than British. They say cozy, esthetic, sizable, theater and draft. We say cosy, aesthetic, sizeable, theatre and draught.

Many Americanisms have become absorbed into British English. We now happily accept words like 'teenage', 'babysitter' and 'commuter'. But the potential for massive misunderstanding still remains. In New York, as part of a toney night out, you may go to an upscale eatery for a dish of zucchini, garbanzo beans and cilantro, cooked in a skillet. In London, a classy night out might include a visit to a posh restaurant for courgette, chickpeas and coriander cooked in a frying pan.

Anglicized words

Words from many other languages have become accepted into British usage and require no special treatment. Juggernaut and bungalow began their life in India. Café, restaurant and rendezvous all come from France. Some words have retained their accent. It is therefore correct, for example, to write cliché, not cliche.

George Orwell, a master of clear and lucid prose, believed 'that you should never use a foreign phrase if you can think of an everyday English equivalent'. If in your letter you mention that you 'felt a frisson of schadenfreude from a mélange of outré haute couture' you may give an impression of being just slightly affected. As Miss Piggy used to say, 'Pretentious? Moi?'

Quotation marks

Nowadays put direct quotations inside single quotes. For example: He said, 'Listen to me.' If the speaker is reporting what someone else has said, then it's double quotes inside single quotes. For example: He replied, 'When I spoke to her, she said, "You listen to me!" ' Which style you use is, of course, a matter for you.

Hyphens

Words like today, midday and no one are never hyphenated. Fractions, two-thirds, four-fifths and so on are written with a hyphen. Be careful of those occasions when using a hyphen changes the meaning of what you are trying to say. There is a big difference between buying a little-used car and buying a little used car.

Beyond that, it is almost impossible to lay down hard and fast rules for when the hyphen should be used. Nonconformist and nonplussed are written without a hyphen. Non-combatant, non-existent, non-payment and non-violent are written with one. If in doubt, look the word up in a good dictionary.

Typos (or literals)

The troble with typos is that they force you to take your eye off the bal. It is very hard to cuncentrate if the eye, and therefore the drain, is consantly being distracted.

Worse, they can fatally undermine your message. Have a look at the following extract. It is taken from *One Hundred Firsts*, a beautifully produced glossy book published to celebrate the centenary of the Savoy Hotel.

Sadly, the diary editors of the broadsheet newspapers fell down on one typo. Instead of positive coverage of a glamorous celebration, much of the publicity focused on one amusing mistake and the impact of the book as a marketing tool was greatly reduced. See if you can spot the typo:

> 83. First hotel to honour the Everest conquerors Hillary and Tensing by flying champagne and caviar out to them in 1953.

84.   In 1967, Lady Whitemore, the Swedish-born wife of British racing driver Sir John Whitemore, became the first woman to die in The Savoy restaurant wearing trousers.

Let's hope she didn't set a trend.

Appendix 2

Making the most of information technology

by David Roth

Many managers and professionals still regard information technology with suspicion or disdain. In fact, when it comes to writing and transmitting letters, the computer can be a powerful ally. This appendix sets out to explain the various ways in which you might benefit from an informed use of IT.

Buying a personal computer

Despite all the brand names you may come across in computer magazines or computer shops, there are essentially only two types of personal computer. These are the IBM-compatible (the PC) and the Apple Macintosh (the Mac).

The PCs are called IBM-compatibles for historic reasons. IBM were the mass marketers of the desktop PC as we now know it and were therefore the people who set the industry standards that most of the other companies followed. Today there are thousands of companies that manufacture IBM-compatible machines other than IBM. Indeed the world market leader for PCs is a company that did not even exist in those early years –

Compaq. It is they who are now setting the industry standards, together with Intel, the leading designer and manufacturer of the main Central Processor Unit (CPU) and Microsoft, the world's largest software company. Despite this, the computer industry still refers to the PC as 'IBM-compatible'.

The other type is the 'Apple Mac'. In many ways Apple was the inventor of the PC, and without Apple there may not have been a mass-market PC industry.

To say that Apple had a revolutionary effect on the development of the PC is a massive understatement. To many, Apple was not just a computer but a helper and friend. However, they went their own individual way and developed a computer system that was not compatible with the IBM model. In many ways it was technically superior to, and easier to use than, the PC. They were devastatingly outmarketed and, although they have an extremely loyal group of users, they command a very small share of the market.

Today some 95 per cent of all personal computers are IBM-compatible. That means that, no matter who makes them, they are able to run any software designed for an IBM PC. Thus it is not surprising that most software written is for the PC. It is probably unwise to consider buying an Apple Mac, not because it is technically deficient but for compatibility and future proofing reasons. So this appendix will concentrate on IBM-compatibles.

Almost any IBM-compatible computer you buy today will come with the operating software called Microsoft Windows. There have been a number of versions of this; the previous one being called Windows® 95 and the current one Windows® 98. This will be followed by a change of name near the turn of the century with a new version, to be called Windows NT. This re-branding is to bring Microsoft's offering in the consumer/small business and large business sectors (where NT is already established) under one name. Windows is the computer operating system and allows you, through a series of icons, to control your PC's functions and start your computer programs, and makes attaching printers and other devices to your PC easier. Think of an operating system as the link between your computer and anything you put in it and anything you attach to it.

Do you need a brand name?

There has been an explosion in the number of 'no name' brands in the PC market place over the past few years. Because of the standardization of components it is relatively easy for anyone who knows what they are doing to assemble a quality PC at a lower price that can compete with the big brands. However, unless you are competent in the field it is best not to buy on the basis of price alone. The real value in putting together a PC is in the way that all the parts communicate with each other, and this is something that the bigger brands spend much time and resource perfecting. It is also worth paying extra for a PC that comes from a company with a good reputation and that offers quality after sales service and support.

Measurements

The computer industry has developed its own unique measurement system. You do not need to know about it in detail, but a little knowledge will help you understand some of the jargon used to describe the computer's parts and why some PCs are more expensive than others:

- data is measured in units of 'bits' and 'bytes'
- a bit is one-eighth of a byte
- a byte contains eight bits, and is the amount of storage needed for the computer to store a single character, for example the letter 'r'
- a kilobyte is a thousand bytes, well 1 024 to be exact
- a megabyte is just over a million bytes (1 048 576)
- and finally a gigabyte is a little over a billion bytes.

Assessing your computer needs

If you want to use a PC simply for word processing, sending and receiving e-mail, doing simple data work on a spreadsheet and tracking your personal finances, choose a standard PC.

If your needs include regular 'number crunching' and data base analysis then the key to selecting the PC that will best suit you will lie in its speed and memory. The golden rule is to get as much as you can afford.

If you are choosing a multimedia PC then the task is a little more daunting, as playing games and editing images puts the greatest strain on the resources of your PC. Knowing what to look for will help you to make sense of the abundant choice that you will have.

The PC system

A PC system that will help you write letters is made up of a combination of individual elements: the central processing unit, hard disc, the monitor, modem, printer and software. Each has its specific part to play, and the correct combination will give you your ideal computer system.

The CPU

The 'Central Processing Unit', the CPU, is the part of the computer that does all the main computing or processing work. The CPUs that were used in the original IBM PCs were called '8088'. As they improved in performance subsequent generations were called '80286', '80386' and '80486' respectively, often shortened to '386' and '486'. A company called Intel developed a commanding worldwide lead in the design and manufacture of these processing chips.

Principally because of the legal difficulties of registering a number as a trademark, the next generation of CPUs developed by Intel were given not a number but the name 'Pentium'. Further developments have been called 'Pentium Pro' and 'Pentium Pro MMX'. Although Intel is the industry leader other companies such as AMD, Cyrix and IBM make 'clone' Intel processors.

What is MMX?
MMX brings to the Pentium chip multimedia capabilities that were only available before with an addition to the CPU chip called an accelerator. An MMX understands standard Pentium

instructions, but adds 57 new instructions specifically tailored towards getting better multimedia performance such as video, 3-D and sound.

Operations that previously required several or even hundreds of commands can now be carried out with just one. This speeds up the processor considerably – but you will need software especially developed for the MMX to see more than just small performance gains.

MMX performs best in the games arena, which relies heavily on fast 3-D graphics and realistic sounds. This is less true for word processing, spreadsheets and databases, which are not multimedia-intensive. Although they will benefit slightly they will show fewer performance gains on a PC with an MMX than on a Pentium.

Does it have to be Intel Inside?
Intel have spent a fortune attempting to convince us all that the only thing that matters about a PC is whether it has an Intel CPU processor – 'Intel Inside'. While it is undoubtedly true that Intel has had a significant effect on the market and have been pioneers in processor development, Intel CPUs are not the only ones that can be used in a PC to achieve excellent results. If you buy a PC from an established manufacturer such as Compaq, IBM or Dell you do not really have to worry whether the CPU is from Intel or from another manufacturer.

More than just the CPU's name
Just knowing the processor's name is only part of the information you will need in order to assess its true performance. A vital ingredient to know is its speed, as it is the combination of this and the type of processor that determines how fast your PC's optimum operating speed will be.

Processor speed is measured in megahertz (MHz) and the higher the number the faster the speed will be. So a 486 running at 100 MHz will be about as fast as a Pentium running at 60 MHz, though much slower than a Pentium running at 150 MHz.

Do not buy a PC unless it is at least a Pentium or Pentium class with a speed of 266 MHz plus and, if it is for multimedia, an MMX.

Secondary cache

Just as important as the chip speed is the Level 2 or secondary cache. The computer's main random access memory cannot supply data fast enough to keep a high-speed processor busy, so a small amount of very speedy and more expensive memory is used as a buffer between main memory and the CPU. A fast Pentium will need at least 256 kilobytes (K) of cache to perform to its full potential. But eventually the law of diminishing returns will apply and the benefits you get per extra K will become negligible.

RAM

Another important factor in optimizing your PC's performance is the amount of Random Access Memory (RAM) that it has. The computer's RAM is where it keeps the programs and data that you have open and available for use at any time. Of course with large programs everything cannot be loaded into RAM so the program will load the most often used parts into RAM and pull from the disc other parts as and when needed. When you turn your computer off everything held in your PC's RAM disappears. In general the more RAM you have the more programs and data you can have open and the faster your PC will work.

If you have insufficient RAM it is likely that your PC will slow down considerably, especially when using heavy graphic-intensive and multimedia programs or when you have a few programs open at the same time.

Despite what you might hear, Windows® 95/98 does need a minimum of 16 MB of RAM to run well. When you add on to that the requirements of multimedia, 32 MB of RAM should be your starting point. You can add additional RAM on to your PC but every PC has its maximum, which you should find out about in your buying process.

Hard disc

The hard disc is made of magnetic material and keeps the data once the PC has been turned off. It is the size of the hard disc that will determine how much data you will be able to store on your computer. The smallest drives installed in today's PCs are 1 GB. You may think that you will never fill that up. But in practice you never have enough disc space. Once you have your operating system (Windows® 95/98), your programs and a good collection of saved pictures and sound files, you will very quickly have gobbled up all your available storage space. So it is wise to get a 3 GB drive or bigger.

After size, speed is the other element of a hard disc that you should consider. The speed in which the hard disc stores and retrieves your data can have an impact on your system's overall performance. Look for a hard disc that is rated 12 ms (milliseconds) or lower (with this measure the lower the better).

Need more space?
In time you will probably find your hard disc full of data that you do not use often but want to keep.

Perhaps the cheapest solution is installing a tape back-up drive. However, these devices are slow and while tapes are good for copying (backing up) and storage, if you want to use the data stored on it you do need to load it back on to your computer's hard disc again.

As an answer to this common dilemma an entire industry has recently sprung up offering what are called 'removable storage drives'.

These drives are similar in structure to the PC's hard disc but the discs that actually hold the data can be removed. With these devices you acquire a virtually unlimited amount of storage space. Once one disc becomes full, you just take it out and put in another.

The main types of removable drives are soft magnetic discs which can hold up to 100 MB of data, like the very popular Iomega's Zip drive and the LS-120 to be found on many Compaq PCs. Their disadvantage is that they are slower than the standard hard drive to read data. The other type uses rigid discs such as Syquest Technology's Flyer and Iomega's Jaz drive. These can store a massive 1/2 GB of data and are as fast as your PC's hard drive. However, the drives are expensive and so are the discs.

CD-ROM

Almost every PC that you can buy will come with a CD-ROM drive (Compact Disc Read Only Memory). A compact disc can hold vast amounts of data, either text, graphics or hi-fi stereo sound. CD-ROM hold 650 MB of data, which is equivalent to about 250 000 pages of text. The speed of a CD-ROM is important. The faster it spins the faster the data can be taken off the disc and used by the PC's processor. The first CD-ROM drives transferred data at 150 kB per second. The speed then doubled to 300 kB, then quad speed at 600 kB. Today, CD-ROMs have surpassed ten times their original transfer rate. The number of times faster than the first 150 kB per second is the way the speed of CD-ROMs are denominated – for example 4x, 6x, 8x. For full-motion video, at least quad speed is required, but now most new PCs have CD-ROMs of 16x. The more sophisticated the multimedia the faster you will want your CD-ROM to be.

What you can have in a desktop PC cannot all be crammed into the small space of a portable so, inevitably, deciding to buy a portable rather than a desktop will involve some sort of compromise. The main trade-off areas are weight, desktop functionality and price. If you want portability so as to be able to work anywhere on the move then choose a portable. If portability is not critical, you can get more for your initial money by buying a desktop and it will be much more flexible and cost-effective should you wish to upgrade it later on.

What to look for in a portable

Keyboard

Some portable manufacturers make weight and size savings by reducing the dimensions of the keyboard and consequently the distance between the keys. If you are a good touch typist or have slightly larger than average-sized fingers you could find that your typing mistakes increase.

Screen

The most important factor is the screen display. The main screen technologies are passive matrix (DSTN) and active matrix (TFT). Active matrix technology gives you a brighter screen and consequently deeper colours. Unlike passive matrix they can also be viewed at almost any angle. Active matrix technology is however more expensive and tougher on battery life.

It is best to acquire as large a display as you can afford.

Weight

If you intend to carry your machine rather than transport it in the back of your car, remember, the lighter the better. When you do your comparative weight calculations do not forget to include the weight of the mains adapter and a spare battery, as the chances are that you will be carrying these around as well and they can add up to a considerable extra weight.

Battery life

A portable is only useful away from the mains if its battery life is long enough. The life of the battery depends on a number of factors:

- *the type of material* the battery is made of – lithium is best, followed by nickel-metal-hydride
- *the type of work you are doing* on your portable – material of great graphic intensity uses up more battery power than simple word processing
- *how well it recharges* – to improve your batteries' life you should discharge the battery completely and recharge it from empty to ensure a full battery charge.

Expandability

Because not everything can be packed into it, the portable that you buy should have two slots for what is known as a PCMCIA card. They enable you to slot in credit card-sized expansion cards which will let you attach all sorts of additional devices, for example a modem. Additionally, though most have these as standard, be sure to check that there is a place to plug in an external monitor, printer, mouse and keyboard, preferably directly into the back of the portable or via a docking station.

Pointing device

Which type of pointing device (the mouse equivalent) is best for you is a matter of personal preference. The three most common types are:

- *a track ball*, where you move the cursor by rolling a fixed ball with your thumb, or index finger
- *a touch pad*, where you control the cursor by moving your finger across a flat surface
- *a nipple*, which looks like the blunt end of a pencil which sticks up in between the B, H and G keys.

All three take considerable practice to use efficiently. And all are slower to use than a standard mouse.

Peripherals

Having bought your PC, you will find that there is a stream of additional items that can add functionality and performance. These are known in the computer trade as 'peripherals'. They generally need a card that has to be plugged into an expansion slot on your PC. Once you run out of slots it becomes more difficult and expensive to add functionality. So make certain that the PC you are interested in has at least four free slots.

Sound quality

As with hi-fi, the difference between cheap and expensive PC sound quality is easy to hear.

Cheaper PCs usually use a 16 bit SoundBlaster compatible system in conjunction with poor quality yet good-looking speakers.

For a more realistic sound look for a PC with a Wavetable soundcard. For even better sound reproduction think about getting a speaker system with a subwoofer; this gives you base sounds that are missed entirely by cheaper speakers.

Monitors

The computer monitor is really a two-part system, the monitor itself and a display adapter or graphics board that is located inside the computer's 'box'.

Monitor sizes are measured diagonally – 14 inches is the most common size but 17 inches is better. Monitors larger than that cost significantly more and unless you are going to do a mass of very detailed fiddly work are probably not worth the extra.

Monitors vary considerably in quality. The determinant of quality you will see on the screen is its combination of resolution, dot pitch, vertical refresh rate and interlacing.

Resolution

The monitor's resolution is the number of pixels (dots of light) that it can display. The higher the resolution the more detail you will be able to see on the screen. What you want is the minimum of a SVGA (Super VGA) which displays 800×600 (800 pixels horizontally and 600 vertically).

Dot pitch

This is the resolution for each colour pixel. For a 14-inch screen look for a dot pitch of .28. Higher dot pitches do make a difference and you will notice deterioration in the clarity of the screen especially with small type, so do not be seduced by a low price – test out the screen for yourself.

Vertical refresh rate

The name given to the number of times that the screen is redrawn from top to bottom, called a 'pass'. The higher the number the better. If the refresh rate is not high enough it will cause an irritating flicker. A refresh rate of 80 per second – 80 hertz (Hz) – is good.

Interlacing

Interlaced monitors refresh every other line of the screen with each pass. Non-interlaced monitors refresh the entire display with a single pass and are therefore less likely to flicker.

It is worth watching different monitors and working out your own personal preferences. Test both text and graphic display.

Printers

There are two main types of printer, 'laser printers' and 'ink jet printers'. Laser printers are best if what you need is fast, high-volume, high-quality black and white pages. Ink jet printers are slower than laser printers and the quality is not as good, but they are cheaper and have the advantage of being able to print out in colour at an affordable price.

Here are the key factors to look out for when selecting your printer:

Speed
Printers vary considerably in speed and the faster they are the more expensive they are. Assess your likely printing output and match it to the speed of your printer otherwise it could severely slow down your productivity. Printer speeds tend to be classified by the number of pages per minute they can produce. Pages that contain graphics take longer to print than pages of text, and on a laser printer the first page of a document will take longer to print than following pages.

Quality
A printer's quality is principally a function of the number of dots per inch that it prints (dpi). The higher the number the better looking both text and graphics will look. Look for a minimum of 720×720 dpi for an ink jet printer and 600×600 for a laser printer. If you want to use thick paper or overhead transparencies check that the printer can handle it and that these will not get stuck in the machine.

Many four-colour ink jet printers are coming on to the market that have an additional quality setting that you can switch to via the printer's software. This, used in conjunction with a special glossy paper, will produce near photographic quality images. This slows the print speed considerably, but the results are worth it for important illustrations.

Modem

In order to communicate with the outside world and link up through the information superhighway to the Internet, or to communicate using e-mail or fax, you will need a modem. A modem is a device that translates data from your computer into sounds that can then be transmitted over the telephone line to be received by another computer or fax machine anywhere in the world where they convert it back into data again.

Here are the key factors:

Speed
The speed of the modem is what determines the maximum speed at which you will be able to send and receive data. If all you are going to do is send and receive text then you do not need to be too concerned about speed. But text is only one element in today's highly graphic, voice- and video-intensive superhighway world.

Modem speed is measured in the number of bits per second (bps) it can transmit. Do not consider a modem of less than 56 000 bps. Try and get the fastest modem available. As well as being less frustrating it can also save you money on your on-line service bills and telephone call charges. The faster the modem the less time it will take you to transmit and receive your data or download files. As on-line services charge by the amount of time you are connected a faster modem will help save you money.

There are two places where the modem can be attached to the PC, 'internally' and 'externally'. Internal modems fit into the expansion slots inside your PC or, on a portable, via the PCMCIA card. An external modem is like an additional box that has its own power supply and attaches to the PC via a cable to the serial port. Each has their relative advantages and disadvantages. Internal modems are cheaper and do not take up additional space, but if you are short of slots could take up valuable resources. External modems can be easily moved from machine to machine and, due to the series of lights, it is easier for you to tell what the modem is doing.

Most multimedia machines will probably come with an internal modem already installed. If you get the choice to upgrade it for the fastest one available my advice is to take it.

Be sure to get a modem that has fax capabilities. This will allow you to fax directly from your computer and receive faxes directly to your computer, which you can then print out.

The new generations of modems also come with telephone answering capabilities. This gives you the option of setting up individual mailboxes for different people.

Although having a fax modem is a great convenience it does not completely replace a conventional fax machine. You cannot use it to send anything that is not in your computer such as a page of a magazine or a hand-written note and most PCs need to be switched on to receive any incoming faxes. Some computers have a sleep mode, which allows the PC to be in a state of semi-consciousness ready to be woken up by an incoming fax. This mode helps conserve power, cuts down on the heat generated and the noise of the fan which can be considerable.

E-mail

E-mail is one of the facilities of the Internet and is currently used by more than 35 million people. The widespread use of e-mail has spawned a new description of conventional mail – 'snail mail'. E-mail allows almost instantaneous sending and receiving of messages, documents, pictures, video and sound anywhere across the world. Anything that can be converted into a digital form can be transmitted via e-mail.

An e-mail message is made up of two main parts:

The header

The main components of this are:

- 'from' – where the message came from
- 'to' – who the message is for
- 'date' – date and time sent
- 'copy' – who is to receive a copy of the message
- and 'subject' – what the message is about.

The body

This is the main text of the message. You can put anything in here you like ranging from one word to an entire book. You could also attach a computer file of a word processing document, spreadsheet, presentation, picture, sound or video file or any kind of combination.

How does e-mail work?

Suppose I wanted to send e-mail to my colleague, Bill, in New York. First I would call up my Internet access provider and, using my e-mail program, type a message and then put Bill's e-mail address in the address space. All addresses are made up of two elements. The server it is being sent to (the domain) and the name of the person. Think of the domain as being the street name and the person the house number. For example:

  yourname@domain.com

or in my example's case:

  Bill@Bigcompany.com

Then I would hit the 'send' button on the e-mail software option.

What happens next is a simple electronic transmission of information. My message is sent to my Internet Service Provider (ISP), which then forwards it to the destination domain. In this case Bill works for a large company that has its own connection to the Internet so he does not need to use a third party ISP. The destination domain is found by looking at the last part of the e-mail address – in this case 'Bigcompany' – which is located in New York. It is rapidly sent across the Atlantic via the phone line to Bill. The next time Bill checks his e-mail he can read my message with his own computer's e-mail program.

Using e-mail, the message arrives in seconds rather than taking five or six days via the post. As well as getting there fast it is also cheap, costing no more in phone charges than the local call to my ISP. This is not only much cheaper than a fax but has much more versatility in being able to send files.

How do I get connected to the information superhighway?

What do you need in order to be able to send e-mails? In addition to your PC you will need an ISP, a web browser and a connection software, a telephone line, a modem and, if you want to send and receive e-mails, an e-mail address.

Telephone line

A standard telephone connection is fine. Only if you are going to make very heavy use of the Internet or will be uploading and downloading very large graphic files is it worth considering a special Integrated Services Digital Network (ISDN) line. This line has a higher bandwidth that allows more data to be sent and received per second. But remember you only get the benefit if the person you are connected to also has ISDN facilities.

A web browser

There are only two real options you have for a web browser – Netscape Navigator or Microsoft's Internet Explorer. The competition between them is fierce and both allow you to do similar things.

Internet Service Provider (ISP)

To link up to the Internet you need to link up with a computer/server that has a connection directly into what is called the backbone of the Internet. If you are a large organization making heavy use of the World Wide Web and e-mail it makes sense to have your own connection. Otherwise it is cheaper, and a great deal easier, to link up via a third party and share the costs with other like-minded people. In order to service this growing group of people a new breed of company has sprung up, the Internet Service Provider, known as ISPs. There are two general categories of ISPs each with their own individual merits.

On-Line Services
In this group CompuServe/America Online are the main players. As well as providing access to the World Wide Web, they offer the advantages of a more simple process of connection, a vast amount of proprietary information that you will not find on the Internet and a much more structured filing system, which is therefore easier to find your way around. On average their access to the Internet tends to be slower (although this is improving) and as they offer more than just access to the Internet also charge a bit more. This is either a flat rate for unlimited usage or a lesser amount for five hours per month with additional charges for each hour or part of an hour after that. Most services offer a local number to dial into which will help keep your phone bills down.

Internet only Service Providers
This second group gives you access to the Internet and e-mail facilities, but some provide added services such as the ability to have your own web home page or set of pages. On the whole they charge a flat monthly fee for unlimited usage. This is slightly cheaper if you pay for 12 months in advance (only worth doing once you have tried out the service for a few months and are happy with it).

Which ISP should you choose?
There is fierce competition so you could just look for the cheapest. But price is not everything. Take care in your choice of ISP as you will find that price and performance generally go hand in hand. So, the lowest charge is no consolation if every time you try to get connected the number you dial is constantly busy, especially during peak hours. Here are the points you should check while researching which ISP you should sign with.

Does your ISP have enough capacity to be reliable? First and foremost you need a reliable, uninterrupted service. This does not happen by chance. Your chosen ISP must have enough physical infrastructure to handle the calls that your make and those of the other subscribers. As a general rule there should be no more than ten subscribers per modem at the ISP's site.

How many hours of connection time will you get for your

monthly fee and is there a free or reduced cost trial period? Is the call you will have to make to get connected to the ISP a local call? If it is not this could add considerably to the total cost of your Internet connection.

Do you get easy to use 'Getting Started' software? You will need this to get your computer to talk to your ISP, send and receive e-mails and surf the Net.

If you want a business account check these factors as well.

Do you get access to a free Web page and address as part of your standard monthly fee? How advanced can the Web pages be? Some ISPs cannot provide all of the features that make your web site interesting, up-to-date and effective for your needs. Can you get a full Web site hosting service? Your business may want a tailor-made name for its Web site (for example, www.yourcompany.com), Web site 'click' reports giving data on how many people are visiting your site, and to be able to take orders on-line using secure credit card transactions.

How much technical support will be available to you? Your business may not have the time or expertise to deal with some of the trials of using the Internet and maintaining an exciting and effective Web site. But, irrespective of whether you will be using your ISP for business or pleasure make sure they offer a technical support telephone line where you can talk to a real human voice rather than a recorded one and at the time of day convenient to you.

Whichever ISP you choose you will get a unique e-mail address with which you can send and receive e-mail and data to anyone on the superhighway.

If you are just thinking about getting connected you will be pleased to know that it is much easier than it was. The Web is becoming so popular that during peak periods at evenings and weekends it is often difficult to get connected. Equally the speed differs at different times of the day. The more users a site has the slower it can become. If the sites you are particularly interested in are located in the USA, as a general rule you will find it quicker to access them in the morning before the States has woken up.

What word processor should I get?

Your PC would be virtually useless if it did not have software, the programs that allow you to do word processing, spreadsheet work or play games. Not so long ago it was difficult for the letterwriter to decide which word processor or presentation package to get. There were so many to choose from, each with their own way of doing things. Choosing was an ordeal. Today word processors and other home and office programs are all packaged up into what are called 'suites' combining word processing, spreadsheet, presentation, finance and communications.

The number of companies offering these products has consolidated into three main players: Corel's 'WordPerfect Suite', Lotus's 'SmartSuite for Windows', and Microsoft's 'Office'. All these offer a full business suite and a smaller and less expensive home/small business version. They are all impressive. Companies have at last understood that it is important for the program to adapt to the way the user wants to use it rather than the other way round. They have also made it much easier to use data generated in one program in another. They offer good help features, sometimes even offering them to you before you realize that you are in need. Once you understand how to get around one part of the program the other parts are easy as they all follow the same layout and principles.

All this has not come without cost. The programs are enormous, devour disc space of anything between 80 MB and 200 MB and are still growing fast. Our expectation of what the word processers now need to do has gone sky high. We expect them to be desktop publishers, e-mail editors, and Web page designers as well.

Which one you choose is now really a matter of personal preference.

A few simple rules to help you keep ahead

Modem and graphic speed shouldn't be skimped on, so, even if you feel that a slower component is sufficient for your current needs, it is well worth sparing a thought for the future. If the past is anything to go by programs will become even bigger and you will need greater speed for them to work at their optimum. So, buying the fastest CPU, modem and video card available that you can afford will delay the need for you to upgrade.

Do not be tempted to upgrade unless it will give you access to added functionality and content that your current PC does not give. If you are only going to save a few nanoseconds with a faster processor, resist the temptation and wait.

In most cases, it makes more sense to pass your system down to your family (or an unsuspecting colleague) every few years rather than upgrading piecemeal. Although it is possible to upgrade your processor it is not really worth it. In the end it is no cheaper than buying a new system, and you'll still have the same older components around your new processor.

It often takes several months for the software and peripherals that take advantage of the latest enhancements to come on to the market. So you can wait for the prices to come down without sacrificing extra functionality. If you buy a PC that has been out for a few months you will still get the productive years without the cutting edge prices.

Do not buy a new PC just before a major innovation, such as a new operating system launch or a generation leap of processors.

Making yourself comfortable

If you follow a few basic rules you can make your time on the PC even more enjoyable by being comfortable.

Although static electricity is not the problem it once was, it is still worth trying to reduce the amount of static generated around your PC. A good quality comfortable chair is a necessity. Get a chair that is designed to be worked in for long periods of time. Make sure that both the seat height and the backrest can be fully adjusted. Try it out before you buy.

The right desk height is very important – a desk that is too high can give you arm strain. Your best height guide is that your elbows should be higher than your wrists while you type.

Make sure that the place where you work has plenty of light. The type of light will also make a difference. Defused light is best as you want to avoid glare on your screen that might give you eye strain. A movable light source is a good idea.

The monitor should be up at eye level when you are sitting up straight. You do not want to be constantly bending over to view the screen.

Take regular breaks in order to avoid general strain to your eyes and wrists.

The Health and Safety Executive have issued guidelines as to the correct position to sit in and lighting conditions that are a statutory requirement for employers. There is also now a legal requirement for employers to make regular eye testing and spectacles available to employees who regularly have to use a computer monitor for any length of time.

What happens when it goes wrong?

In general the two types of support you might need when things go wrong with your PC are mechanical and software support. What seems to you to be a problem with the PC could very easily be to do with the software and not with the physical machine. The stores are very keen to sell extended warranties. This is not surprising as they make a great deal of money out of them. Extended warranties might give you peace of mind but, as a rule, any mechanical or electronic defect that might cause your PC to break is likely to show itself within the first year (while under the manufacturer's guarantee) or not at all.

Given the rate of change in the PC market you might even want a new one before your current one breaks! Remember that extended warranties do not cover routine questions or software-related issues, the most common problems of Internet communications, or getting your printer to work properly. Most manufacturers offer some type of customer support. Make sure that the type of support is available when you need it. Some do not operate late at night or at weekends. Be prepared to pay extra for good quality support even with the big brands.

Enjoy ...

Owning and using a PC should be both productive and fun.
There is very little physical damage you can do to your PC by
just clicking the mouse and experimenting. Make sure you have
a copy of your data and you can just relax and enjoy your
access to the digital age.

Index